ETHICS IN

Information Technology

George Reynolds

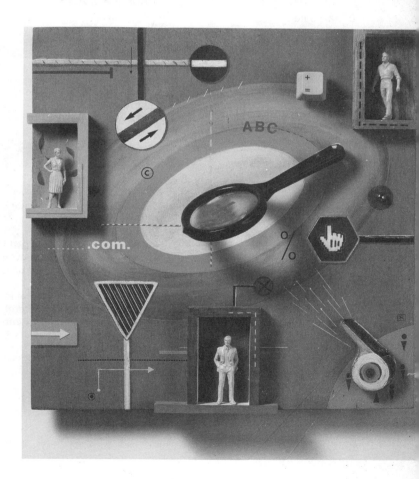

THOMSON

COURSE TECHNOLOGY

Australia • Canada • Mexico • Singapore • Spain • United Kingdom • United States

THOMSON

COURSE TECHNOLOGY

Ethics in Information Technology

George W. Reynolds

Product Manager:
Tricia Boyle

Executive Editor:
Jennifer Locke

Development Editor:
Jeanne Herring

Associate Product Manager:
Janet Aras

Editorial Assistant:
Christy Urban

Associate Production Manager:
Christine Spillett

Cover Illustrator:
Rakefet Kenaan

Cover Designer:
Betsy Young

Composition/Prepress:
GEX Publishing Services

Manufacturing Coordinator:
Denise Powers

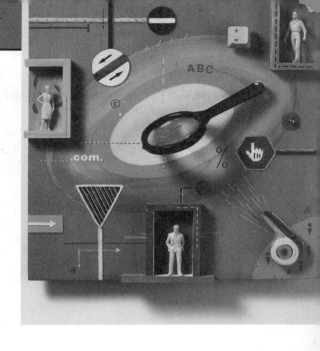

BRIEF Contents

TABLE of Contents

PREFACE

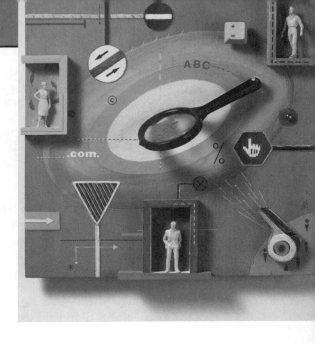

Ethics in Information Technology fills a void of practical business information for business managers and IT professionals. The typical Introduction to Information Systems text devotes one chapter to ethics and IT and provides little practical advice. Such limited coverage does not really meet the needs of business managers and IT professionals—the people primarily responsible for addressing ethical issues in the workplace. What is missing is an examination of the different ethical situations that arise in the realm of information technology and, where appropriate, practical advice for addressing these issues. Although just eight chapters and less that 300 pages long, *Ethics in Information Technology* has enough substance that an instructor could easily use it in a full semester course in computer ethics. *Ethics in Information Technology* could also be used as an additional reading supplement for courses such as Introduction to Management Information Systems, Principles of Information Technology, Managerial Perspective of Information Technology, Computer Security, E-Commerce, and so on.

ORGANIZATION

There are eight chapters in the book, each addressing a different aspect of ethics in information technology. Chapter 1, "An Overview of Ethics", provides an introduction to ethics, ethics in business, and the relevance of the discussion of ethics to information technology. Some philosophical approaches to ethical decisionmaking are discussed, and a suggested model process for ethical decisionmaking is presented.

Chapter 2, "Ethics for IT Professionals and IT Users", explains the importance of ethics to an IT professional in his or her professional relationships, and the role that certification and licensing have (or don't have) in legitimizing professional standards. The significance of IT professional organizations and their codes of ethics are also emphasized in the chapter and are included in the corresponding appendices at the end of the book.

Chapter 3, "Computer and Internet Crime", describes the types of ethical decisions IT professionals must make, and the business needs that must be balanced, in dealing with security issues. It provides a useful classification of the kinds of computer crime and computer crime perpetrators that exist. In addition, the chapter provides information about how to manage security vulnerabilities and respond to specific security incidents in order to fix problems quickly and improve security measures going forward.

Chapter 4, "Privacy", covers the issue of privacy and the effect that the use of information technology has on privacy rights. Although there has been much legislation about privacy rights over the years, little of it affects private businesses specifically. Information that businesses gather about individuals using information technology can be useful to obtain or keep customers (or to monitor employees), but there is concern about how much information can be gathered, who it can be shared with, how the information is gathered in the first place, and how it is used. This is true not just in business but in law enforcement and government as well.

Chapter 5, "Freedom of Expression", addresses the topic of freedom of expression and the issues raised by the growing use of the Internet as a means for that expression. The First Amendment of the United States provides extensive protection of speech of all kinds, even if offensive or annoying. However, the ease with which Internet users can communicate, and the fact that they can do it anonymously, poses many dilemmas for people and groups who are negatively affected by that communication. In addition, attempts to control access to Internet content that is unsuitable for children, or unsuitable for a business environment, has caused the development of laws and filtering software; in both cases, the effectiveness of the solution has been limited.

Chapter 6, "Intellectual Property", defines intellectual property and explains how technology is and is not protected by copyright, patent, and trade secret laws. Key intellectual property issues relevant to ethics in IT are discussed, including reverse engineering of software, software license legislation, competitive intelligence gathering, and cybersquatting.

Chapter 7, "Software Development", provides a thorough discussion of the software development process and the importance of software quality. Many issues are covered that software manufacturers must consider when deciding "how good is good enough?" with regard to their software products, particularly when that software is safety critical and its failure can cause loss of human life. These include deciding what risks are being taken, the extent of quality assurance to use, and standards that may be adopted to achieve quality goals.

Chapter 8, "Employer/Employee Issues", addresses two significant topics having to do with the relationship of employers to those who work for them. The first topic is the use of contingent workers, contractors who are United States citizens as well as H-1B workers and employees of offshore outsourcing companies. The chapter discusses the reasons companies might decide to use contingent workers and the ethical issues that are raised when they are making those decisions. The second topic is whistle blowing. Discussed are the risks and protections of whistle blowing, the ethical decisions that must be made by whistle blowers, and a process for approaching blowing the whistle when someone decides action must be taken against his or her own company. Both topics in the chapter have ethical, social, and economic implications and should be of great interest to students of ethics.

PEDAGOGY

Ethics in Information Technology employs a variety of pedagogical features to enrich the learning experience and provide interest for the instructor and student.

Opening Quotation. Each chapter starts with an opening quotation to stimulate interest in the material and set the stage for the chapters.

Vignette. A brief real world example of a company illustrating the issues to be discussed in the chapter is presented to peak the reader's interest.

Focus Questions. Carefully crafted focus questions follow the vignette to further highlight specific topics to be covered in the chapter.

Key Terms. Key terms are placed in bold in the text. All key terms are defined in the glossary at the end of the book.

Additional Features

Two additional features appear in each chapter of the text to maintain the reader's interest and motivation. Each chapter includes a Legal Overview and a Manager's Checklist. These features are designed to support the goals and themes of the text and the content of the specific chapter. They can also be used to generate class discussion.

- **Legal Overview.** Provides a brief summary of an important aspect of the law as it applies to ethics and information technology.
- **A Manager's Checklist.** Provides a highly practical and useful checklist of questions that address the key points that must be considered when making a business decision.

End-of-Chapter Material

To help students retain key concepts and to expand their understanding of important IT concepts and relationships, a Summary, Review Questions, Discussion Questions, What Would You Do Exercises, and Case Projects are included at the end of every chapter.

Summary. Each chapter includes a summary of the key issues raised in the chapter. The items in the summary relate back to the Focus Questions posed at the start of the chapter.

Review Questions. Directly linked to the text, these questions reinforce the key concepts and ideas within each chapter.

Discussion Questions. Picking up where the review questions leave off, these more open-ended questions help instructors generate class discussion to move students beyond the concepts to explore the numerous aspects of ethics in information technology.

What Would You Do? These ethics exercises present real-to-life dilemmas that can be used as the basis of student exercises to involve the student in the critical thinking process using the principles presented in the text.

Case Projects. Two real-world cases are provided for each chapter. These cases reinforce important ethical principles and IT concepts and show how real companies have addressed ethical issues associated with IT. Critical thinking questions focus students on the key issues of each case and ask them to consider and apply the concepts presented in the chapter.

ABOUT THE AUTHOR

George Reynolds brings a wealth of computer and industrial experience to this project, with more than 30 years' experience working in government, institutional, and commercial IS organizations. He has also authored fourteen texts and is an adjunct professor at the University of Cincinnati, teaching a distance learning course on Legal and Ethical Issues of IT.

TEACHING TOOLS

The following supplemental materials are available when this book is used in a classroom setting. All of the teaching tools available with this book are provided to the instructor on a single CD-ROM. Some of these materials may also be found on the Course Technology Web site at **www.course.com**.

Electronic Instructor's Manual. The Instructor's Manual that accompanies this textbook includes:

- Additional instructional material to assist in class preparation, including suggestions for lecture topics.
- Solutions to all end-of-chapter exercises.

ExamView®. This textbook is accompanied by ExamView, a powerful testing software package that allows instructors to create and administer printed, computer (LAN-based), and Internet exams. ExamView includes hundreds of questions that correspond to the topics covered in this text, enabling students to generate detailed study guides that include page references for further review. The computer-based and Internet testing components allow students to take exams at their computers, and also save the instructor time by grading each exam automatically.

PowerPoint Presentations. This book comes with Microsoft PowerPoint slides for each chapter. These are included as a teaching aid for classroom presentation, to make available to students on the network for chapter review, or to be printed for classroom distribution. Instructors can add their own slides for additional topics they introduce to the class.

Distance Learning. Course Technology is proud to present online courses in WebCT and Blackboard, as well as MyCourse 2.0, Course Technology's own course enhancement tool, to provide the most complete and dynamic learning experience possible. When you add online content to one of your courses, you're adding a lot: self tests, lecture notes, a gradebook, and, most of all, a gateway to the twenty-first century's most important information resource. Instructors are encouraged to make the most of your course, both online and offline. For more information on how to bring distance learning to your course, contact your local Course Technology sales representative.

ACKNOWLEDGMENTS

I would like to thank a number of people who helped greatly in the creation of this book: Christine Spillett, Production Editor, for smoothing out the production process; Jeanne Herring, Developmental Editor, for her incredible insight and constructive feedback; Tricia Boyle, Product Manager, for overseeing and directing this effort; Nikki Brainard, for her feedback and encouragement as she read early versions of the text; and my many students at the University of Cincinnati who provided excellent ideas and constructive feedback on draft versions of the text. I would especially like to thank Jennifer Locke, our Managing Editor, for her faith in me and in this project. In addition, I would like to thank a marvelous set of reviewers who offered many useful suggestions. They are as follows:

John P. Buerck, Ph.D., Saint Louis University; Jeffrey P. Corcoran, Lasell College; Robert Fulkerth, Ph.D., Golden Gate University; Howard Sundwall, Communications Test Design, Inc.; Pete Weaver, Oklahoma City Community College; Donna J. Werner, Saint Louis University.

Last of all, thanks to my family for all your support and allowing me to take the time to do this.

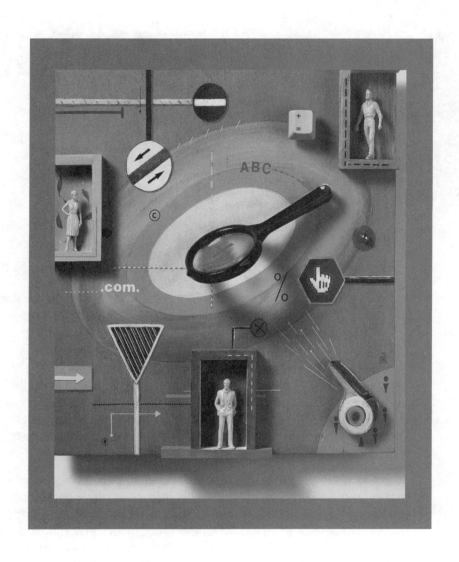

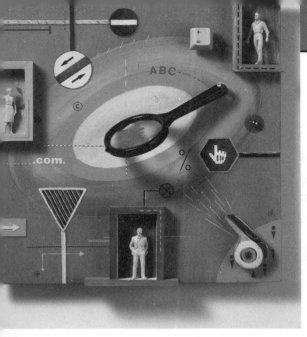

Chapter 1

An Overview of ETHICS

"This above all: to thine own self be true, and it must follow, as the night the day, thou canst not then be false to any man."
— Polonius, a character in Shakespeare's *Hamlet*, discussing the importance of moral integrity with his son, Laertes.

The Enron financial debacle in the latter half of 2001 thrust corporate accounting and auditing into the spotlight, subsequently bringing closer attention to other companies that have been forced to issue restatements of previous earnings reports. One such company is Peregrine Systems, a San Diego firm that develops software for businesses to track their assets, such as IT equipment and fleets of vehicles.

On May 5, 2002, the firm announced that its board of directors had authorized an internal investigation into potential accounting inaccuracies uncovered by its KPMG auditors. Based on preliminary information, certain transactions that involved revenue recognition irregularities amounting to $100 million—one-fifth of the firm's 2001 revenue—were called into question. Additionally, the board announced the resignations of its chairman of the board, chief executive officer, and chief financial officer. Peregrine's stock plummeted to $.89—a huge markdown from its split-adjusted high of $79.50 in March 2000. A firm that *Business Week* named one of the 100 best performing IT companies for 2001 and whose customers include 92 percent of the Fortune 500 companies had fallen on hard times.

Some of the company's investors wondered how revenue could have been overstated. Actually, there are many different and legal ways companies can report revenue figures. However, investors tend to sue first and look for facts later—more than a dozen class-action shareholder lawsuits were filed against the company.

Adapted from Chris Gaither, "Peregrine Shares Hit Skids on Disclosure of an Audit," *The New York Times* on the Web, May 7, 2002, accessed at www.nytimes.com; "Peregrine Systems Announces Internal Accounting Investigation; CEO and CFO Resign," Press Release May 6, 2002 from Peregrine Web site at www.peregrine.com; and Kim Peterson, "Numbers Get Massaged in Corporate Accounting," *San Diego Union Tribune*, May 12, 2002, accessed at www.signonsandiego.com.

As you read this chapter, consider the following questions:

1. What are ethics and why is it important to act in ways that are consistent with a code of principles?

2. Why are business ethics becoming increasingly important?

3. What actions are corporations taking to reduce business ethics risks?

4. Why are corporations interested in fostering good business ethics?

5. What approach can you take to ethical decision making?

6. What trends have increased the risk of negative impact due to the unethical use of information technology?

WHAT ARE ETHICS?

Each society forms a set of rules that establishes the boundaries of generally accepted behavior. Often, the rules are expressed in statements about what people should or should not do. These rules fit together to form the **moral code** by which that society lives. Unfortunately, there are often contradictions among the different rules, and you can become uncertain about which rule you should follow. For instance, if you witness a friend copy someone else's answers while taking an exam, you might be caught in a conflict between loyalty to your friend and the value of telling the truth. Sometimes, the rules do not seem to cover new situations, and you must determine how to apply the existing rules or develop new ones. You may strongly support personal privacy, but in a time when employers are tracking employee e-mail and Internet usage, what rules do you think are acceptable to govern appropriate use of company resources?

The term **morality** refers to social conventions about right and wrong that are so widely shared that they are the basis for an established common consensus. However, one's view of what is moral may vary by age, cultural group, ethnic background, religion, and gender. There is widespread agreement on the immorality of serious crimes such as murder, theft, and arson. However, certain behaviors that are accepted in one culture might be unacceptable in another. For example, in the United States, it is perfectly acceptable to place one's elderly parents in a managed care facility to provide them with a high level of care and attention in their declining years. In most Middle Eastern countries, however, one's elderly parents would never be placed in such a facility (even if they were available); they remain at home and are cared for by other family members. As another example, attitudes toward the illegal copying of software (software piracy) range from strong opposition to acceptance as a standard approach to business. In 1998, 38 percent of all software in circulation worldwide was pirated, but in Vietnam, 97 percent of the software used was pirated. In both Indonesia and Russia, the figure was 92 percent.[1] Lastly, even within the same society, there can be strong disagreements over important moral issues—in the United States, for example, abortion, the death penalty for criminals, and gun control are all continuously debated topics where both sides feel their arguments are on solid moral ground.

Definition of Ethics

Ethics are beliefs regarding right and wrong behavior. Ethical behavior refers to behavior that conforms to generally accepted social norms. On many key issues, the identification of ethical behavior is very clear. Almost everyone would agree that lying and cheating are examples of unethical behavior. On many other issues, what constitutes ethical behavior is subject to individual interpretation. For example, most people would not steal an umbrella that belongs to someone else; however, upon finding an umbrella under a seat in a theater, one person might think it's ethical to keep it, while another might think the only ethical course of action is to turn it in to the theater's "lost and found." An individual's interpretation of ethical behavior is strongly influenced by a combination of family influences, life experiences, education, religious beliefs, personal values, and peer influences.

As we mature, we develop habits that allow us to perform complicated actions or learn new things—walking, riding a bike, or writing the letters of our alphabet—without having to think about them consciously. We also develop habits that make it easier for us to choose between what is generally accepted by our society as good or bad. **Virtues** are habits that incline us to do what is acceptable and **vices** are habits that incline us to do what is unacceptable. Fairness, generosity, honesty, and loyalty are examples of virtues. Vanity, greed, envy, and anger are examples of vices. A person's virtues and vices help define that person's **value system**, the complex scheme of moral values by which one chooses to live. Thus, a person's virtues and vices help determine that person's interpretation of what is ethical and help govern that person's behavior.

The Importance of Integrity

Your moral principles are statements of what you believe to be rules of right conduct. As a child, you may have been taught a principle along the lines of "Don't lie, cheat, or steal or have anything to do with those who do." As an adult making your own, much more complex decisions, you often reflect on your principles when you consider what you ought to do in different situations: Is it okay to lie to protect someone's feelings? Can you keep the extra $10 you received when the cashier mistook your $10 bill for a $20 bill? Should you intervene with a coworker who seems to have an alcohol or chemical dependency problem? Is it okay to exaggerate your work experience on a résumé? Can you "cut some corners" on a project to meet a tight deadline?

If a person acts with **integrity**, that person acts in ways that are consistent with his or her own code of principles. Indeed, integrity is one of the cornerstones of ethical behavior. One approach to acting with integrity is to extend to all persons the same respect and consideration that you yourself desire. Unfortunately, it is difficult to act in a consistent manner. One opportunity for inconsistent ethical behavior arises when you find yourself in a situation in which there is a conflict between your standards. Suppose, for example, that you believe it is important for your career to do as your employer requests, and you also believe that you should be fairly compensated for work that you do. Now, imagine that, due to budget constraints, your employer insists that you not report some hours of overtime that you have worked on a project. The situation creates a conflict between your moral standards. You can do as your employer requests, or you can insist on being fairly compensated, but you cannot do both. In this situation, you may be forced to act in a manner inconsistent with one of your principles and to act with an apparent lack of integrity.

Another form of inconsistency emerges if we apply our moral standards differently to different people or to different situations. To be consistent and act with integrity, we must apply the same moral standards to one person or one situation that we apply to another. For example, you might consider it morally acceptable to tell a friend a "little white lie" to spare the person some pain or embarrassment, but would you consider lying to a work colleague or customer about a business issue to avoid unpleasantness?

The remainder of this chapter provides an introduction to ethics in the business world. It discusses the importance of ethics in business, outlines the actions that businesses are taking to reduce ethics risks, points out that good ethics are not always good business, provides advice on creating an ethical work environment, and suggests a model of ethical decision making. The chapter concludes with a discussion of ethics and information technology and provides a brief overview of the remainder of the text.

ETHICS IN THE BUSINESS WORLD

Risk is the product of multiplying the likelihood of a negative event happening by the impact of such an event happening. Thus, if the likelihood of an event happening is high and the potential negative impact is large, the risk is considered great. Ethics have come to the top of the business agenda because the risks associated with inappropriate behavior have increased both in their likelihood and in their potential negative impact on the organization.

Several trends have led to the increase in the risk of unethical behavior. First, the increased globalization of organizations has created a much more complex work environment that spans diverse societies and cultures and that makes it much more difficult to apply principles and codes of ethics consistently. For example, numerous United States companies have garnered negative publicity for moving operations requiring heavy manual labor to third-world countries where the employees work under conditions that would not be acceptable in most developed parts of the world.

Second, employees, shareholders, and regulatory agencies are increasingly sensitive to issues such as violations of accounting standards, failures to disclose substantial changes in business conditions to investors, nonconformance with required health and safety practices, and production of unsafe or substandard products. Their increasing vigilance raises the risk of financial loss (through lawsuit, loss of business, or loss of business value) to businesses that do not foster ethical practices or that run afoul of required standards. For example, energy company Enron's accounting practices hid the real value of the firm, and in late 2001 the company was forced to file for the largest corporate bankruptcy in United States history. In 2002, Schering-Plough was fined $500 million by the Food and Drug Administration because of its repeated failure over the years to fix problems in manufacturing dozens of drugs at four of its factories.[2] In the same year, the General Electric Company and IBM both changed their financial disclosure practices following pressure from investors for clearer and fuller reporting.[3]

These trends have led to an increased need to focus on business ethics.

Why Fostering Good Business Ethics Is Important

There are five reasons why corporations need to promote an ethical work environment by encouraging employees to act ethically when making business decisions and by supporting them when they do so:

1. To protect the organization and its employees from legal action
2. To create an organization that operates consistently
3. To produce good business
4. To avoid unfavorable publicity
5. To gain the goodwill of the community

Protect the Corporation and Its Employees from Legal Actions

In 1991, the United States Justice Department published sentencing guidelines that suggested more lenient treatment for convicted corporate executives if their companies had established ethics programs. In 1995, Delaware (a favorite state for incorporation by many companies) courts warned that corporate directors could be held personally liable for subordinates' wrongdoing if they had failed to establish programs to ensure compliance with the law. For criminal violations, the federal government has set sentencing guidelines that encourage the establishment of effective ethics programs and that impose stunning fines for violations. However, fines can be lowered by as much as 80 percent if the organization has implemented an ethics management program and cooperates with authorities.[4] Yet, the dishonest or unethical actions of a single employee can negatively affect an entire organization. Read the Legal Overview to learn more about bribery, one of the most common white-collar crimes committed by employees.

BRIBERY

Bribery involves providing money, property, favors, or anything else of value to someone in business or government in order to obtain a business advantage. An obvious example is a software supplier that offers money to another company's employee to get business from that company. This type of bribe is often referred to as a "kickback" or "payoff." Bribery is one of the most prevalent forms of white-collar crime. The person offering a bribe commits the crime of bribery when the bribe is offered. The person receiving the offer of a bribe is guilty of the crime of bribery when he or she accepts the bribe.

The United States Foreign Corrupt Practices Act (FCPA) makes it a crime to bribe a foreign official, a foreign political party official, or a candidate for foreign political office. The act applies to any United States citizen or company or to any company with shares listed on any United States stock exchange. However, the payment of a bribe is not a crime if the payment was lawful under the laws of the foreign country in which it was paid. Penalties for violating the FCPA are quite severe. For corporations, it is up to $2 million per violation. Individuals may be fined up to $100,000 and imprisoned for up to five years.

The FCPA also requires corporations to meet its accounting standards by having an adequate system of internal accounting controls, including maintaining books and records that accurately and fairly reflect their transactions. The goal of these standards is to prevent companies from using "slush funds" or other means to disguise payments to foreign officials. A firm's business practices and its accounting information systems are frequently audited by both internal and outside auditors to ensure that they meet these standards.

The FCPA permits facilitating payments that are payments made for "routine government actions," such as obtaining permits or licenses; processing visas; providing police protection; providing phone services, power, or water supplies; or facilitating actions of a similar nature. Thus, it is permissible under the FCPA to pay an official to perform some official function faster (for example, to speed customs clearance) but not to make a different substantive decision (for example, to award business to one's firm).

In some countries, gifts are an essential part of doing business. In fact, there are countries where it would be considered rude not to bring a present to the initial business meeting. In the United States, a gift might take the form of free tickets to a sporting event by a sales rep for a personnel agency that wants to get on your company's list of preferred suppliers (or stay on that list). At what point does a gift become a bribe? Who decides?

The key distinguishing factor is that no gift should be hidden. A gift may be considered a bribe if it is not declared. As a result, most companies require that all gifts be declared and that those above a token value must be declined. Some

(continued)

companies have a policy of pooling the gifts received by its employees, auctioning them, and giving the proceeds to charity.

When it comes to distinguishing between bribes and gifts, the perceptions of the donor and recipient can differ. The recipient may believe that what is received is a gift that in no way obligates the recipient to the donor. This applies particularly when the gift is a benefit other than cash. The donor's intentions, however, might be very different. Table 1-1 will help you distinguish between a gift and a bribe.

Table 1-1   Distinguishing between a bribe and a gift

Bribes	Gifts
Are made in secret as they are neither legally nor morally acceptable	Are made openly and publicly as a gesture of friendship or goodwill
Are often made indirectly through a third party	Are made directly from donor to recipient
Encourage an obligation for the recipient to alter his or her behavior in some way favorable to the donor	Come with no expectation of a future favor for the donor

Adapted from: "Dealing with Bribery and Corruption — A Management Primer," the Royal Dutch Shell Group Web page at www.shell.com/ accessed on March 30, 2001 and Henry R. Cheeseman, *Contemporary Business Law, 3rd edition*: (Upper Saddle River, New Jersey: Prentice-Hall, 2000), p.148.

Create an Organization That Operates Consistently

Organizations develop and abide by values that will create an organization that operates consistently and in a manner that meets the needs of its many stakeholders—shareholders, employees, customers, suppliers, and the community. Special emphasis is given to those values needed to address current issues in the workplace and to tackle the organization's strengths, weaknesses, opportunities, and threats. Although each company's value system is different, many organizations have the following values in common:

- Operate with honesty and integrity, staying true to what we believe
- Operate according to our standards of ethical conduct—both in words and action ("walk the talk")
- Treat our colleagues, customers, and consumers as we want to be treated
- Strive to be the best at what matters most to our company
- Accept personal responsibility for our actions
- Value diversity
- Make fact-based, principle-based decisions

Produce Good Business

In most cases, good ethics mean good business. Companies that produce safe and effective products avoid costly product recalls and lawsuits. Companies that provide excellent customer service maintain their existing customers rather than lose them to competitors. Companies that develop and maintain strong employee relations suffer less from high employee turnover and low employee morale. Suppliers and other business partners often treat companies that operate in a fair and ethical manner in a similar manner. Thus, good ethics can improve the profitability of the organization.

It is also true that bad ethics can lead to bad business results. For example, many employees' negative attitudes and behaviors arise from a difference between the employees' personal values and the values stated or implied by the organization's actions. In such cases, employees often act to defend themselves against anticipated punitive actions and/or to retaliate against the organization for how they have been treated. Such an environment destroys employee commitment to organizational goals and objectives, creates low morale, fosters poor performance, leads to low employee involvement in corporate improvement initiatives, and builds employee indifference to the needs of the organization.

Avoid Unfavorable Publicity

The public reputation of a company strongly influences the value assigned to its stock by shareholders, how consumers regard its products and services, the degree of oversight it receives from government agencies, and the amount of support and cooperation it receives from business partners. Thus, one reason companies are motivated to build a strong ethics program is to avoid negative publicity. If the organization is perceived as operating in an ethical manner, customers, business partners, shareholders, consumer advocates, financial institutions, and regulatory bodies will regard it more favorably.

Those companies that are perceived as operating other than 100 percent ethically quite often suffer negative consequences. As we saw in the opening vignette, when a public company is discovered to have misreported its earnings, its stock price can be negatively affected.

Gain the Goodwill of the Community

Although organizations are primarily run to earn profits or provide services to their clients, they also have some basic social responsibilities that they owe to society. Many corporations recognize these responsibilities and make a serious effort to fulfill them. Often, these responsibilities are declared in a formal statement of a company's principles or beliefs. Some of the types of corporate activities that show social responsibility include making contributions to charitable organizations and nonprofit institutions, providing benefits for employees in excess of any legal requirements, and choosing an economic opportunity that is judged to be more socially desirable over a more profitable alternative.

The "goodwill" that socially responsible activities create can make it easier for corporations to conduct their businesses. For example, a company that is known for the good treatment of its employees will find it easier to compete for the best job candidates. On the other hand, companies that are viewed as harming their community may be placed at a disadvantage. For example, a firm that pollutes the air and water may find that the negative publicity generated by its actions reduces sales and makes it difficult to work closely with some business partners.

Improving Corporate Ethics

The risks of unethical behavior are increasing, and there are many benefits to acting ethically in business. Thus, the improvement of business ethics is becoming increasingly important. Here are some of the actions corporations are taking to reduce business ethics risks.

Appointment of a Corporate Ethics Officer

The **corporate ethics officer** is a senior-level manager responsible for improving the ethical behavior of the members of an organization. Typically this is done by establishing an environment that encourages ethical decision-making through some or all of the actions described in this section of the chapter. The role of the corporate ethics officer has become increasingly common. However, simply putting someone in the position of corporate ethics officer will not automatically lead to improved ethics; it will take a lot of hard work and effort to establish and provide ongoing support for an organizational ethics program.

Ethical Standards Set by Board of Directors

A 1999 Conference Board study of 124 companies in 22 countries found that a corporation's board of directors is now much more involved in the creation of ethical standards. Although only 21 percent of the policies in existence in 1987 showed board involvement, participation had increased to 41 percent in 1991 and to 78 percent by 2000.[5] Gaining the attention of the board of directors shows that an increased level of importance is placed on ethical standards.

Establish a Corporate Code of Conduct

A **code of conduct** is a guide that highlights an organization's key ethical issues and identifies the overarching values and principles that are important to the organization and that can help in decision making. It frequently includes a set of formal, written statements about the purpose of the organization, its values, and the principles that guide its employees' actions. The code of conduct helps ensure that employees abide by the law, follow necessary regulations, and behave in an ethical manner.

For a code of conduct to be accepted company wide, it must be formulated with employee participation and be fully endorsed by the organization's leadership. In addition, it must be easily accessible by employees, shareholders, business partners, and the public. The application of a code of conduct to decision making must be continually emphasized as an important part of an organization's culture. Lastly, breaches in the code of conduct must be identified and treated as appropriate so that the relevance of the code of conduct is not undermined.

Establishing a corporate code of conduct is an important step for any company, and a growing number of companies have done this. In May 2001, *Business Ethics* magazine rated United States-based, publicly held companies based on a statistical analysis of corporate service to seven stakeholder groups—employees, customers, community, minorities, domestic shareholders, the environment, and overseas stakeholders. The top five companies were Procter & Gamble, Hewlett Packard, Fannie Mae, Motorola, and IBM.[6] The code of conduct from top-rated Procter & Gamble is summarized in Table 1-2.

Table 1-2   Procter & Gamble Code of Conduct

Procter & Gamble Purpose

P&G people are committed to serving consumers and achieving leadership results through principle-based decisions and actions. We will provide products and services of superior quality and value that improve the lives of the world's consumers. As a result, consumers will reward us with leadership sales, profit and value creation, allowing our people, our shareholders, and the communities in which we live and work to prosper.

Procter & Gamble Core Values

P&G People We attract and recruit the finest people in the world.
Leadership We are all leaders in our area of responsibility, with a deep commitment to deliver leadership results.
Ownership We accept personal accountability to meet the business needs, improve our systems, and help others improve their effectiveness.
Integrity We always try to do the right thing.
Passion for Winning We are determined to be the best at doing what matters most.
Trust We respect our P&G colleagues, customers, and consumers and treat them as we want to be treated.

Procter & Gamble Principles

We show respect for all individuals.
The interests of the company and the individual are inseparable.
We are strategically focused in our work.
Innovation is the cornerstone of our success.
We are externally focused.
We value personal mastery.
We seek to be the best.
Mutual interdependency is a way of life.

Source: The Procter & Gamble Company

Conduct Social Audits

An increasing number of companies are conducting social audits of their policies and practices. In a **social audit**, companies identify ethical lapses committed in the past and take actions to avoid similar mistakes in the future. For example, Mattel, the toy maker, embarked on a social audit led by an advisory panel made up entirely of company outsiders. The panel spent a year creating workplace standards to ensure that none of its overseas plants were exploiting workers. Although the social audit found none that were, it did uncover accounting infractions and air-quality abuses at some factories and made sure the local managers corrected them.[7]

Require Employees to Take Ethics Training

The ancient Greek philosophers believed that personal beliefs regarding right and wrong behavior can be taught. Today most psychologists agree with them. Lawrence Kohlberg, the late Harvard psychologist, found that many factors stimulate a person's moral development, but one of the most crucial factors is education. Other researchers have repeatedly supported these findings—people can continue their moral development through further education that involves critical thinking and examination of contemporary issues.[8]

Thus, a company's code of conduct must be promoted and continually communicated within the organization, from top to bottom. To do so, organizations should present employees with examples of how the code of conduct is applied in real life. One approach to doing this is through a comprehensive ethics education program that encourages employees to act responsibly and ethically. Often such programs are presented in small workshop formats that include employees applying the organization's code of conduct to realistic hypothetical case studies. For example, Procter & Gamble requires all its employees to take a four-hour workshop on the topic of principle-based decision making based on the principles in the corporate code of conduct. Workshop participants must decide the best way to respond to real-life ethical problems, such as the need to give honest and constructive feedback to an employee not meeting expectations. They are also given examples of recent company decisions made using principle-based decision making. Not only does requiring employees to take such courses make employees more aware of the company's code of ethics and how it can be applied in decision making, but also it serves the purpose of demonstrating that the company intends to operate in an ethical manner. The existence of formal training programs can also reduce the company's liability in the event of legal action.

Include Ethical Criteria in Employee Appraisal/Reward Systems

Employees are increasingly being evaluated on their demonstration of qualities and characteristics stated in the corporate code of conduct. For example, in many companies, a portion of an employee's performance evaluation is based on treating others fairly and with respect; operating effectively in a multicultural environment; accepting personal accountability to meet business needs; continually developing oneself and others; and operating openly and honestly with suppliers, customers, and other employees. These factors are strongly considered along with the employee's overall contribution to moving the business ahead, successful completion of projects, and the maintenance of good customer relations.

When Good Ethics Result in Short Term Losses

Operating ethically does not always guarantee business success. Many organizations operating outside of the United States have found that "business as usual" in foreign countries can place them at a significant competitive disadvantage.

A major global telecommunications company placed itself at a significant competitive disadvantage by consistently applying its corporate values to its South American business. Although the organization's code of conduct prohibited the practice of financially "influencing" decision makers on project bids, its competition did not play by the same rules. As a result, the company lost many projects (and therefore millions of dollars in revenues). Senior management argued in favor of integrity and the consistent application of corporate ethics. They reasoned that situational ethics was wrong and that if the practice were ever started, it would quickly run out of control. Their hope was that good ethics was good business in the long term.[9]

Creating an Ethical Work Environment

Most employees want to perform their jobs both successfully and ethically. So why do good employees sometimes make bad ethical choices? It is often due to a highly competitive workplace that places great pressure on its employees. The pressures include aggressive competitors, cut-throat suppliers, unrealistic budgets, minimum quotas, tight deadlines, and bonus incentives for meeting performance goals. In addition, employees may be encouraged to do "whatever it takes" to get the job done. Such an environment can put employees in a situation in which they feel forced to make unethical choices or engage in unethical conduct to meet management's expectations. One should not be surprised if good people do unethical things in such a pressure-packed atmosphere, especially if there are no corporate codes of conduct and no strong examples of senior management practicing ethical behavior. Table 1-3 shows how management's behavior can result in unethical employee behavior. Table 1-4 provides a manager's checklist to assess if you are promoting or working in an ethical workplace. In each case, the preferred answer is "yes."

Table 1-3 How management can affect employees' ethical behavior	
Managerial Behavior That Encourages Unethical Behavior	**Resulting Individual Feelings and Behavior**
Set and hold people accountable for meeting "stretched" goals, quotas, budgets	My boss wants results, not excuses, so I have to cut corners to meet the goals my boss has set.
Fail to provide a corporate code of conduct and operating principles to guide decision-making	Because there are no guidelines, I don't think my action is really wrong or illegal.
Fail to act in an ethical manner and set a poor example for others to follow	I have seen other successful people take unethical actions and not suffer negative repercussions.
Fail to hold people accountable for unethical actions	No one will ever know the difference, or if they do, so what?
When employees are hired, plop a three-inch binder entitled "Corporate Business Ethics, Policies, and Procedures" on their desks. Tell them to "read it when you have time and sign the attached form that says you read and understand the corporate policy."	This is overwhelming. Can't they just give me the essentials? I can never absorb all this.

Table 1-4	Manager's checklist for establishing an ethical work environment

Questions	Yes	No
Does your company have a corporate code of conduct?	——	——
Was the corporate code of conduct developed with broad input from employees at all levels within the organization and does it have their support?	——	——
Is the corporate code of conduct concise and easy to understand, and does it identify those values you need to operate consistently and meet the needs of your various stakeholders?	——	——
Does each employee have easy access to a copy of the corporate code of conduct, and has each signed a document stating that he or she has read it and understood it?	——	——
Do employees participate in annual training to reinforce the values and principles that make up the corporate code of conduct?	——	——
Do you "walk the talk" and set an example by both communicating the corporate code of conduct and actively using it in your decision making?	——	——
Do you evaluate and provide feedback to employees on how they operate with regard to the values and principles in your corporate code of conduct?	——	——
Do you seek feedback from your employees to ensure that the work environment that you create for them does not create situations that are at odds with the corporate code of conduct?	——	——
Do employees believe that you are a fair and just person and do they seek your advice when they see examples of others violating the company's code of conduct?	——	——
Do employees have an avenue, such as an anonymous hotline, for reporting infractions of the code of conduct?	——	——
Are employees aware of sanctions for breaching the code of conduct?	——	——

Ethical Decision-Making

Often in business, the ethically correct course of action is clear and easy to follow, so people act accordingly. Exceptions occur, however, when considerations of ethics come into conflict with the practical demands of business. Dealing with these situations is challenging and risky to one's career. How, exactly, should you think through an ethical issue? What questions should you ask? What factors should you consider? This section will lay out a seven-step approach useful in guiding your ethical decision-making; however, ethical decision-making is not a simple, linear activity. It is important to realize that information gained or a decision made in one step may cause you to go back and revisit previous steps.

Get the Facts

The first step in analyzing any potential issue is simply to get the facts. Many times, innocent situations are turned into unnecessary controversies simply because one does not bother to check the facts. For example, you observe your boss take what appears to be an employment application from an applicant. Your boss throws the application in the trash after the applicant leaves. Such action is a clear violation of your company's policy to treat each applicant with respect and to maintain a record of all applications for one year. You could report your boss to his superiors for failure to follow this policy or you could take a moment to speak directly to your boss about the situation. You might be pleasantly surprised and find out that the situation was not what it appeared to you. Perhaps the "applicant" was actually a salesperson promoting a product for which your company has no use, and the "application" was marketing literature.

Identify the Stakeholders and Their Positions

A **stakeholder** is someone who stands to gain or lose from how a particular situation is resolved. Stakeholders often include others besides those directly involved in an issue. Identifying the stakeholders will help you better understand the impact of your decision and, hopefully, help you make a better decision. Unfortunately, it may also cause you to lose sleep as you worry about how you may affect the lives of others. You may recognize the need to get the stakeholders involved in the decision and thus help gain their support for the recommended course of action. What is at stake for each stakeholder? What does each stakeholder value and what is his or her desired outcome? Do some have a greater stake because they have special needs or because we have special obligations to them? To what degree should they be involved in the decision?

Consider the Consequences of Your Decision

The consequences of decision-making can be viewed from several perspectives. Often, the decision you make will directly impact you. However, you must guard against taking too narrow an approach and simply focusing on what's best for you as an individual. Another perspective is to consider what impact—harms or benefits—your decision will have on all the stakeholders. A third perspective is to consider the impact on the organization. Will your decision help the organization meet its goals and objectives? Yet a fourth perspective is to consider the impact the decision will have on the broader community of other organizations and institutions, ordinary citizens, and the environment. As you view a problem and proposed solution from each of these perspectives, you may gain additional insights that lead you to modify your decision.

Weigh Various Guidelines and Principles

Are there any laws that apply? Certainly you do not want to violate a law that can lead to a fine or imprisonment for yourself or others. If the decision does not have any legal implications, what existing corporate policies or guidelines apply? What guidance does the corporate code of conduct offer? Are there certain personal principles that apply?

Philosophers have developed many different approaches to deal with moral issues. Four of the most common approaches are virtue ethics, utilitarian, fairness, and common good. These theories, summarized in Table 1-5 and discussed in subsequent sections, provide a framework within which decision makers can reflect on the acceptability of actions and evaluate moral judgments. Individuals must find the appropriate balance between all the principles, guidelines, and theories available to help them make decisions.

Virtue ethics approach.   Virtue ethics is a philosophical approach to ethical decision making that focuses on how we ought to behave and how we should think about relationships if we are concerned with our daily life in a community. It does not define an exact formula for ethical decision making. Rather, virtue ethics suggests that, when faced with a complex ethical dilemma, people do either what they are most comfortable doing or what they think a person they admire would do. The assumption is that people will be guided by their virtues to reach the "right" decision. A proponent of virtue ethics believes that a disposition to do the right thing is a more effective guide than following a set of principles and rules and that morality should be something that we do not think about, but merely do out of habit.

Virtue ethics can be applied to the business world by equating the virtues of a good businessperson with those of a good person. However, businesspeople face situations that are peculiar to business, and so they may need certain business-related character traits. For example, honesty and openness in our dealings with others is generally considered to be virtuous; however, for a corporate purchasing manager negotiating a multimillion dollar deal, there may be a practical need to be vague in discussions with competing suppliers.

A problem with the virtue ethics approach is that it doesn't provide much of a guide for action. The definition of "virtuous" cannot be worked out objectively; it depends on the circumstances—you make it up as you go along. For example, bravery is a great virtue in some circumstances; in others, it may be just plain stupid. The right thing to do in a situation will depend on which culture you're in and what the cultural norm dictates.

Utilitarian approach.   This approach to ethical decision-making states that when we have a choice between alternative actions or social policies, we choose the action or policy that has the best overall consequences for all persons directly or indirectly affected. The goal is to find the single greatest good by balancing the interests of all affected parties.

Utilitarianism fits easily with the concept of value in economics and the use of cost-benefit analysis in business. Business managers, legislators, and scientists weigh the resulting benefits and the harm of policies when deciding whether, for example, to invest resources in building a new plant in a foreign country, to enact a new law, or to approve a new prescription drug.

A complication of this approach is that it is often difficult, if not impossible, to measure and compare the values of certain benefits and costs. How do you assign a value to one human's life or to a pristine wildlife environment? Another complication is that it is difficult to predict the full benefits and the harm resulting from our actions.

Fairness approach. This approach to ethical decision-making focuses on how fairly our actions and policies distribute benefits and burdens among those affected by the decision. The guiding principle of this approach is to treat all people the same. However, decisions made with this approach can be influenced by personal biases toward a particular group, and the decision-maker may not even realize it. In addition, there are times when the intended goal of an action or policy (for example, affirmative action, selected tax cuts, and farm subsidies) is to provide benefits to a target group of people, which may seem "unfair" to another group affected by the action.

Common good approach. The common good approach to decision making is based on a vision of society as a community whose members work together to achieve a common set of values and goals. It is based on the principle that the ethical choice is the one that advances the common good. It results in decisions and policies that put in place the social systems, institutions, and environments on which we all depend and in a manner that benefits all people. Examples of such actions and policies would include implementing an effective education system, building a safe and efficient transportation system, and providing accessible and affordable health care.

As with the other approaches to ethical decision-making, there are complications. People clearly have different ideas about what constitutes the common good, thus making it difficult to agree on a common set of values and goals. In addition, maintaining the common good often requires that particular individuals or groups bear costs that are greater than those borne by others (for instance, property owners must pay property taxes to support public school systems while apartment dwellers do not).

Table 1-5   Philosophical theories for ethical decision-making

Approach to Dealing with Moral Issues	Principle
Virtue Ethics Approach	The ethical choice is the one that best reflects moral virtues in ourselves and our communities.
Utilitarian Approach	The ethical choice is the one that produces the greatest excess of benefits over harm.
Fairness Approach	The ethical choice is the one that treats everyone the same and does not show favoritism or discrimination.
Common Good Approach	The ethical choice is the one that advances the common good.

Develop and Evaluate Options

In many cases, it is possible to identify several answers to a complex ethical question. By listing the key principles that apply to the decision, one can usually quickly focus in on the two or three best options. What benefits and what harm will each course of action produce, and which alternative will lead to the best overall consequences? The option chosen should be one that is ethically defensible while at the same time meets the legitimate needs of economic performance and the company's legal obligations.

Review Your Decision

Is this decision consistent with your own personal values as well as those of the organization? How would others (co-workers, stakeholders, business partners, friends, and family) regard you if they knew the facts of the situation and the basis for your decision? Would they see it as right, fair, and good? If you were any one of the stakeholders, would you be able to accept this decision as a fair one?

Evaluate the Results of Your Decision

After the decision has been implemented, monitor the results to see if the desired effect is achieved and observe the impact on those affected by the decision. This will allow you to adjust and improve your decision-making process for future decisions.

ETHICS IN INFORMATION TECHNOLOGY

The growth in the use of the Internet, the ability to capture and store vast amounts of personal data online, and our increasing reliance on information systems in all aspects of our lives have increased the risk of negative impact due to the unethical use of information technology. In the midst of the many information technology breakthroughs of recent years, the importance of ethics and human values has been underemphasized—with a range of consequences. Here are some examples of situations that raise public concern about the ethical use of information technology:

- Today's workers are subject to the monitoring of their e-mail and Internet access while at work, as employers and employees struggle to balance the need of the employer to manage important company assets and employees' work time versus the employees' desire for privacy and self-direction.
- Millions of people have used Napster software to download music at no charge and in apparent violation of copyright laws.
- Robert Hanssen, a career F.B.I. agent, was convicted of providing data from classified databases to Russia.
- DoubleClick, an advertising network that tracks users as they move around the Internet, was sued after it revealed plans to match a mass mailing marketing list with its anonymous database of Internet users, thus revealing the Web users' identities.
- Students around the world have been caught downloading material from the Internet and plagiarizing content for their term papers.
- Hackers engaged in acts of cyberterrorism defaced hundreds of Web sites and left hate messages after a collision between a United States spy plane and a Chinese jet fighter.

This book is based on two fundamental tenets. First, the general public has not realized the critical importance of ethics applied to information technology; too much emphasis has been placed on the technical issues. Information technology is seen as a tool to get a job done, particularly by end users who understand only very narrow slices of ethics and information technology. These slices include virus protection, hacking, and privacy. However, society has become more sensitive to unethical behavior, and so have the courts. It would thus benefit end users and IT professionals to better evaluate situations in a manner consistent with society's expectations. Second, in the corporate world, technical decisions requiring managerial action are often left to the technical experts. General business managers must assume greater responsibility, but to do so they must have the capability to make broad-minded, objective, ethical decisions based on technical savvy, business know-how, and a sense of ethics. They must also do their best to create a working environment where ethical dilemmas can be discussed openly, objectively, and constructively.

Thus the goals of this text are to educate people about the tremendous impact ethical issues play in the successful and secure use of information technology; to motivate people to be mindful of these issues when making business decisions; and to provide tools, approaches, and insights useful in making ethical decisions.

An Overview of This Text

The chapters in this book cover a range of topics relating to ethics in information technology. Chapter 2 discusses the multiple ways that ethics are important to IT professionals and IT users. Chapter 3 addresses computer crime, an area of growing concern to anyone who uses networked information technology. Chapter 4 covers the important issues of personal data privacy and employee monitoring. Chapter 5 deals with the ethical issues raised by Internet communications and freedom of expression. Chapter 6 covers the protection of intellectual property rights through such means as patents, copyrights, and trade secrets. The ethical issues raised by the software development process are addressed in Chapter 7. Lastly, in Chapter 8, the use of "non-traditional" employees, and the ethical dilemmas it can cause, are discussed, as are the implications of whistle-blowing. It is in recognition of the great impact of ethical considerations on these different topic areas that *Ethics in Information Technology* was written.

Summary

1. What are ethics and why is it important to act in ways that are consistent with a code of principles?

   Ethics are beliefs regarding right and wrong behavior. If a person acts with integrity, that person acts in ways that are consistent with his or her own code of principles—integrity is one of the cornerstones of ethical behavior.

2. Why are business ethics becoming increasingly important?

   Business ethics are becoming increasingly important because the risks associated with inappropriate behavior have grown in number, complexity, likelihood, and significance.

3. What actions are corporations taking to reduce business ethics risks?

   Corporations can take a number of actions to reduce business ethics risks: appoint a corporate ethics officer, set ethical standards at a high organizational level, establish a corporate code of conduct, conduct social audits, require employees to take ethics training, and include ethical criteria in employee appraisal systems.

4. Why are corporations interested in fostering good business ethics?

   There are five reasons why corporations are interested in fostering good business ethics: to protect the company and its employees from legal action, to create an organization that operates consistently (because good

ethics can be good business), to avoid negative publicity, and to gain the goodwill of the community. Being ethical, however, is not always a guarantee of business success.

5. What approach can you take to ethical decision making?

   One approach to ethical decision making involves these seven steps: get the facts of the issue, identify the stakeholders and their positions, consider the consequences of your decision, weigh various guidelines and principles, develop and evaluate various options, review your decision, and evaluate the results of your decision. This is not a linear process, and some backtracking and repeating of previous steps may be required.

6. What trends have served to increase the risk of negative impact from the unethical use of information technology?

   The growth in the use of the Internet, the ability to capture and store vast amounts of personal data online, and our increasing reliance on information systems in all aspects of our life have served to increase the risk of unethical uses of information technology. In the midst of our many information technology breakthroughs of recent years, the importance of ethics and human values has been underemphasized—with a range of consequences.

Review Questions

1. Define the word ethics. Define the term value system.

2. Define the term risk. Distinguish between a high-risk and low-risk situation.

3. What trends have led to an increased need for organizations to foster an ethical environment?

4. In what ways do good ethics engender good business?

5. What is the Foreign Corrupt Practices Act (FCPA)? What role do accounting information

systems play in enabling a firm to adhere to this act?

6. How would you distinguish a bribe from a facilitating payment? From a gift?

7. Identify six specific actions corporations are taking to reduce business ethics risks.

8. What are the key elements of a corporate code of conduct? What is its purpose?

9. Identify five reasons why corporations want to take a stronger ethical position than in the past.

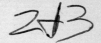

10. Identify six things that managers can do to create an ethical work environment.

11. What are some of the pressures that are placed on employees that can make it difficult for them to perform ethically?

12. Outline and briefly discuss a seven-step approach for ethical decision making.

13. Identify several areas in which the increased use of IT has raised ethical concerns.

Discussion Questions

1. Can you recall a situation in which you had to deal with a conflict in values? What was the situation? What process did you use to resolve this issue?

2. Is every action that is legal also ethical? Can you describe an action that is legal, yet one that you feel would be ethically wrong? Is every ethical action also legal? Is law, not ethics, the only relevant guide that business managers need to consider? Explain.

3. What should managers have done in the case of the major global telecommunications company that could not compete effectively in Latin America? Compromised the company's ethics? Pulled out of the area entirely?

4. This chapter lists four approaches to dealing with moral issues. Identify and briefly summarize each one. Is there one perspective that you believe is the most important? Why or why not?

5. Is it possible for an employee to be successful in the workplace without acting ethically?

6. What are the key elements of an effective corporate-ethics-training program?

What Would You Do?

Use the seven-step approach to ethical decision-making to analyze the following situations and answer the questions that follow.

 1. You are in charge of awarding all personal computer service-related contracts for your employer. In a recent e-mail to the current holder of the service contract for your company, you casually exchanged ideas about home landscaping, your favorite pastime. In your e-mail you stated that you wished you had a few Bradford pear trees in your yard. Upon returning from a vacation, you discover three mature trees in your yard. There is a brief "thank you" note in your mailbox signed by the local sales representative of your company's current personal computer service contractor. You really want the trees, but you certainly didn't mean for the contractor to buy them for you. You suspect that the contractor interpreted your e-mail comment as a hint

that you wanted him to provide you with some trees. You also worry that the contractor still has a copy of your e-mail. If the contractor sent your boss a copy of your e-mail, it might look as if you were trying to get a bribe from the contractor. Can the trees be considered a bribe? What would you do?

2. While mingling with friends at a party, you mention a recent promotion that has placed you in charge of evaluating bids for a large computer hardware contract. A few days later, you and your significant other receive an invitation to dinner at the home of a couple you know slightly who were also at the party. Over cocktails, the conversation turns to the contract you're managing. Your host seems remarkably well-informed about the bidding process and likely bidders. You volunteer information about the potential value of the contract and briefly outline the criteria your firm

An Overview of Ethics

will use to select the winner. At the end of the evening, he surprises you when he reveals that he is a consultant currently working for several of the companies active in this market. Later that night your mind is racing. Did you reveal information that can provide one of the suppliers with a competitive advantage? What are the potential business risks and ethical issues that this situation raises? Should you report this conversation to someone? If so, to whom and what would you say?

3. You are the manager of the customer service desk of a small software manufacturer. The newest addition to your ten-person team is Kelly, a recent college graduate. She is a little overwhelmed by the volume of calls, but is learning fast and doing her best to keep up. Today, as you performed your monthly review of your employee's e-mail, you are surprised to see that Kelly is receiving a number of e-mails from employment agencies. One of the messages says, "Kelly, I'm sorry you don't like your new job. We have lots of job opportunities that I think would better match your interests. Please call me and let's talk further." You're shocked and alarmed. You had no idea she

was so unhappy, and your team desperately needs her help to handle the onslaught of calls generated by the newest release of software. If you're going to lose her, you'll need to find a replacement quickly. You know that Kelly did not intend for you to see this personal e-mail, but you can't ignore what you saw. Should you confront Kelly and demand to know her intentions? Should you avoid any confrontation and simply begin seeking her replacement? What would you do?

4. A coworker calls you at 9 A.M. at work and asks for a favor. He is having a little trouble getting going this morning and will be an hour or so late getting into work. He explains that he has already been late for work twice this month and a third time will cost him four hours pay. He asks you to stop by his cubicle, turn his computer on, and place some papers on the desk so that it looks like he is "in." You have worked on some small projects with this co-worker and gone to lunch together. He seems nice enough and does his share of the work, but you are not sure what to tell him. What would you do?

Cases

1. Missing White House E-mail

E-mail problems in the Clinton White House became public in February, 2000, through a $90 million class-action suit filed by Judicial Watch, a conservative group that had brought multiple lawsuits against the Clinton administration since 1994.

In a sworn affidavit given in February 2000, Sheryl Hall, a computer specialist at the White House, said that she learned that beginning in August 1996, incoming e-mails to the Executive Office of the President were not transferred to computer systems in the White House. This transfer is part of a process called records management. Successful records management would allow the text of the e-mails to be searched in response to

subpoenas and other inquiries. The e-mails involved were those sent to much of the West Wing staff, including the President and the top staff of the President and Mrs. Clinton.

Sheryl Hall went on to testify that because of this problem, when a search of e-mails was done in the White House in response to a subpoena from an independent counsel, e-mails for the time period of November 1996 through at least November 1998 could not be searched. Although the White House officials learned of the records management problem (and its impact on searches in response to subpoenas) in May 1998, the problem was not fixed until after November 1998. As a result, six additional months of incoming e-mails were not records managed so that they could be searched for subpoenas.

The lost e-mails occurred at a time when members of Congress, the Justice Department, and the Office of Independent Counsel had issued subpoenas demanding all relevant White House documents related to campaign funding, the Branch Davidian siege in Waco, Texas, and President Clinton's relationship with Monica Lewinsky. These e-mails were also part of a larger case that originated when Judicial Watch sued the F.B.I. and other defendants after the Clinton administration was given access to hundreds of F.B.I. files on government appointees who served under Presidents Reagan and Bush. Judicial Watch further alleges that the administration threatened the workers who maintained the e-mail system with retribution if they revealed the lack of e-mail backups.

The White House insisted that the error that caused the e-mail problem was entirely unintentional and blamed it on technicians who inadvertently capitalized the name of the mail server designed to store the messages, effectively sending them to a storage server that didn't exist.

According to further testimony from the contractors responsible for the e-mail system to the House Government Reform Committee in March 2000, the problem was first detected in January 1998 by Daniel A. Barry, a member of the White House computer staff. When lawyers asked him to search the server for e-mails related to Lewinsky, Barry found two messages sent to Lewinsky, but he couldn't find the corresponding e-mails from Lewinsky. Barry reported the incident to his supervisor, but he wasn't sure whether the missing e-mails were a minor glitch or an indication of a bigger problem.

In March 2000, White House Counsel Beth Nolan wrote to Representative Dan Burton, chairman of the House Government Reform Committee, that back-up copies of the e-mail were stored on computer tapes, but that reconstructing them would cost between $1 million and $3 million and take as long as two years.

Clinton administration officials have insisted that they didn't realize the scope of the problem, still don't know how many e-mails were affected, and do not know if the e-mails were pertinent to investigations.

Sources: adapted from Todd R. Weiss, "Judge Orders IT Workers to Testify in White House E-mail Case", *Computerworld*, December 15, 2000 accessed at www.computerworld.com.cwi; "White House E-Mail Probe Upheld", *JSOnline Milwaukee Journal Sentinel*, August 25, 2000 accessed at www.jsonline.com; "FBI to Search for White E-Mails", *USA Today* June 30, 2000, accessed at www.usatoday.com/; Timothy W. Maier, "Clinton Created His Own Monster", *Insight Magazine*, June 12, 2000 accessed at www.insightmag.com; John P. Martin, "White House Says Y2K Got in the Way of Investigation into Missing E-Mails", *Fox News*, May 3, 2000 accessed at www.foxnews.com; John P. Martin, "Computer Specialists Tell of Threats In Hearing on Lost White House E-Mails", *Fox News*, March 23, 2000, www.foxnews.com; Sheryl Hall's Affidavit dated February 16, 2000 posted at www.database.townhall.com.

Questions:

1. What is the White House records management process? What is the significance of the records management process?

2. Daniel Barry discovered a problem with the records management process in January 1998, but didn't determine the extent of the problem. What further actions might he have taken to correctly assess the situation while avoiding potential personal repercussions?

3. What could Barry's supervisor have done to identify the extent of the problems earlier? Why might he have failed to follow through on resolving the problem? Can you identify any reasons why there was a six-month delay from the time the problem was finally recognized until it was fixed?

2. McKesson HBOC Accused of Accounting Improprieties

HBO (named for founders Huff, Barrington, and Owens) & Company was formed in November 1974. The company quickly made a name for itself by delivering cost-effective patient information and hospital data collection systems. Its premiere product, MEDPRO, was designed to be the most cost-effective patient information system in the industry. MEDPRO helped hospital administrators track patient admissions, discharges, emergency room registrations, order communications and results reporting, scheduling, and data collection.

The company went public in June 1981 under the NASDAQ stock symbol HBOC. HBOC's fast growth in sales and profits made it a favorite of investors throughout the 1990s. Its stock rose more than a hundredfold between October 1990 and October 1998. Much of its growth came through acquisitions of other companies.

On January 12, 1999, McKesson Corporation, one of the largest distributors of prescription drugs in the United States, completed a merger with HBO & Company by exchanging 177 million shares of McKesson common stock for all shares of common stock of HBOC. The company was renamed McKesson HBOC, Inc. At the time, the two companies had a combined market value of more than $23 billion.

Just three months after the merger, McKesson HBOC Inc. announced that its auditors had discovered accounting irregularities at HBOC during a routine annual review. The irregularities were uncovered when McKesson's accounting firm, Deloitte & Touche, mailed a survey to several clients asking for the amount of goods and services they had actually purchased from the company. Several of the amounts returned by clients did not match what HBOC had recorded. As a result, McKesson HBOC, Inc. would have to restate earnings for the last four quarters. When the restatement of earnings was announced in April 1999, shares of the company plunged from $65 to $34 in a single day.

Some believed that certain HBOC managers, seeking to ensure that HBOC would meet or beat analysts' expectations for sales and profits, had used several innovative approaches to report the company's financial results. Throughout 1998, it was alleged that HBOC allowed more than a dozen hospitals to buy HBOC software or services with conditional "side letters" that enabled the hospitals to back out of the sales. It was further alleged that these side letters were not shared with the auditors and that the associated sales were reported as complete, violating accounting rules. In at least one case, a hospital canceled a purchase HBOC had booked.

Additional allegations were made that in order to bolster its results, HBOC agreed to questionable sales with two other large computer companies. In September 1998, two days before the end of HBOC's quarter and just weeks before HBOC and McKesson agreed to merge, HBOC agreed to buy $74 million in software from Computer Associates, supposedly for resale. In return, Computer Associates, a maker of software used by big companies, bought $30 million in HBOC software, also supposedly for resale. The deals were split into separate contracts, neither of which made any reference to the other. In the two years following its purchase of HBOC's software, Computer Associates had neither made use of nor distributed any of the $30 million in HBOC software products, according to a court indictment. Similarly, HBOC had sold only a fraction of the $74 million in Computer Associates software it bought.

In yet another damaging allegation, in March 1999, McKesson HBOC, Inc. and Data General, a computer hardware maker based in Massachusetts, agreed to a similar deal. Data General disclosed to auditors that what appeared to be a simple $20 million purchase of HBOC's software had a side agreement that essentially ensured that Data General would never have to pay for it.

Investors and analysts had no idea the problems uncovered would be so costly. McKesson first said it had found $42 million in sales from its HBOC unit that had been improperly booked. Eventually, $327 million in overstated revenue and $191 million in overstated income over three years (1997–1999) were uncovered.

In one of the drug industry's largest corporate shakeups, the board of McKesson HBOC, Inc. ousted some of its top executives over the accounting irregularities. In July 1999, the United States Attorney's Office for the Northern District of California and the Securities and Exchange Commission (SEC) started investigations. In addition, 53 class action lawsuits, 3 derivative actions, and 2 individual actions were filed against the company and certain current or former officers and directors of the company. Two former top executives of HBOC were indicted in July 2000, accused of costing investors more than $9 billion in one of the largest financial frauds in American history. As of this writing, legal proceedings to determine the guilt or innocence of the two executives continue; if convicted, they face up to 10 years in jail and a $1 million dollar fine.

Sources: adapted from David J. Morrow, "The Markets: Market Place; McKesson to Restate Earnings for 4 Quarters and Stock Falls 48%," *New York Times on the Web*, April 29, 1999 accessed at www.nytimes.com; Alex Berenson, "Two Ex-Executives Are Indicted In Fraud Case," *New York Times on the Web*, September 29, 2000 accessed at www.nytimes.com; "Suit Contends Board Of McKesson Knew Of Problems at HBO," *New York Times on the Web*, November 15, 2000 accessed at www.nytimes.com; McKesson HBOC, Inc. Form 10-Q filed with Securities and Exchange Commission for quarter ended December 31, 2000 and accessed at Edgar Online at www.edgar-online.com; McKesson HBOC, Inc. Form 10-K Annual Report for the year ended March 31, 1999 filed with Securities and Exchange Commission and accessed at Edgar Online at www.edgar-online.com.

Questions:

1. Make a list of the specific people and parties hurt by the use of non-standard accounting practices at HBOC. For each entity, identify the harm suffered.
2. As you read this case, was there a clear point at which ethical wrongdoings became legal wrongdoings? If so, at what point?
3. Can you identify possible causes and motivations for HBOC managers to use non-standard accounting procedures to report increased revenues and earnings?

Endnotes

[1] Microsoft Software Piracy Web page – Privacy Basics – Worldwide Piracy accessed at www.microsoft.com on April 29, 2001.

[2] Melody Petersen, "Drug Maker to Pay $500 Million Fine for Factory Lapses," *New York Times on the Web*, May 18, 2002, accessed at www.nytimes.com.

[3] "Financial Update—Clarity in Reporting Gets Hip," *InformationWeek*, February 25, 2002, accessed at www.informationweek.com.

[4] Sreenath Sreenivasan, "Taking In the Sites; Finding Business Ethics on and About the Internet," *New York Times on the Web*, March 23, 1998 accessed at www.nytimes.com.

[5] Amy Zipkin, "Getting Religion On Corporate Ethics; A Scourge of Scandals Leaves Its Mark," *New York Times on the Web*, October 18, 2000 accessed at www.nytimes.com.

[6] Philip Johansson, "The 100 Best Corporate Citizens for 2001," *Business Ethics Magazine,* May 2001 accessed at www.business-ethics.com.

[7] Amy Zipkin, "Getting Religion On Corporate Ethics; A Scourge of Scandals Leaves Its Mark," *New York Times on the Web*, October 18, 2000 accessed at www.nytimes.com.

[8] "Can Ethics Be Taught?", Santa Clara University Markkula Center for Applied Ethics accessed at www.scu.edu on March 19, 2001.

[9] Sreenath Sreenivasan, "Taking In the Sites; Finding Business Ethics on and About the Internet," *New York Times on the Web*, March 23, 1998 accessed at www.nytimes.com.

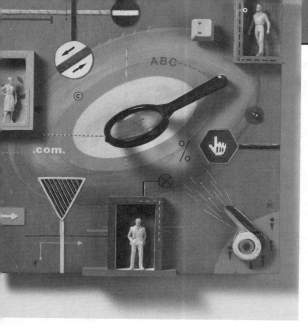

Chapter 2

ETHICS
for IT
Professionals
and IT Users

"I feel that licensing is and will indefinitely continue to be inappropriate for software engineers."
— Fred Brooks, author of the classic project management book, *The Mythical Man-Month* (1975)

"...the time for instituting licensing software engineers is already here."
— Dave Parnas, Director of Software Engineering Program, McMaster University, Ontario, Canada[1]

Increasing incidents of poor software quality and reliability have raised concerns about the software development methods and the qualifications of the people involved. State governments, licensing bodies, and individuals have suggested that software engineers should be licensed to protect the public interest. The Texas Professional Engineers Licensing Board approached the Association for Computing Machinery (ACM) in 1998 to help define performance criteria for software engineering for its state licensing examinations. Initially the ACM agreed to take on this task even though some members of the ACM had reservations about whether licensing software engineers was in the best interests of the field of computing and the public. After several meetings and much review by a blue-ribbon panel of computing professionals, the ACM Council in a May 1999 meeting stated that it opposed the licensing of software engineers because it believes that such licensing is premature and would not effectively address the problems of software quality and reliability.

As you read this chapter, consider the following questions:

1. What are the key characteristics that distinguish a professional from other kinds of workers and what is the role of an IT professional?

2. What are the various professional relationships that must be managed by the IT professional and what are the key ethical issues that can arise in each?

3. How do codes of ethics, professional organizations, certification, and licensing affect the ethical behavior of IT professionals?

4. What are the key tenets of four different codes of ethics that provide guidance for IT professionals?

5. What are the common ethical issues that face IT users?

6. What approaches can be taken to support the ethical practices of IT users?

IT PROFESSIONALS

According to Webster's dictionary, a **profession** is a calling requiring specialized knowledge and often long and intensive academic preparation. The United States has adopted labor laws and regulations that require a more precise definition of what is meant by a "professional" employee. The United States Code of Federal Regulations [29 CFR Sec. 541.3] defines a person "employed in a professional capacity" as one who meets these four criteria:

1. One's primary duties consist of the performance of work requiring knowledge of an advanced type in a field of science or learning customarily acquired by a prolonged course of specialized intellectual instruction and study or work.
2. One's instruction, study, or work is original and creative in character in a recognized field of artistic endeavor and the result of which depends primarily on the invention, imagination, or talent of the employee.
3. One's work requires the consistent exercise of discretion and judgment on its performance.
4. One's work is predominately intellectual and varied in character and is of such character that the output produced or the result accomplished cannot be standardized in relation to a given period of time.

In other words, a professional is someone who requires advanced training and experience, must exercise discretion and judgment in the course of his or her work, and whose work cannot be standardized. Many people would add that they expect a professional to contribute to society, to participate in a life-long training program (both formal and informal), to keep abreast of developments in their field, and to help develop other professionals. In addition, many professional roles carry with them special rights and special responsibilities. Doctors, for example, are allowed to prescribe drugs, perform surgery, and request confidential patient information. They are also expected to help individuals who are hurt in accidents or emergencies. Accountants, lawyers, priests, and licensed engineers also have special rights and responsibilities.

Definition of an IT Professional

Given the definition of "professional," there are many business workers whose duties, background and training, and work could qualify them to be classified as professionals. These people include marketing analysts, financial consultants, and IT specialists. A partial list of IT specialists includes programmers, systems analysts, software engineers, database administrators, Local Area Network (LAN) administrators, and chief information officers (CIOs). It could be argued that not every IT role requires "knowledge of an advanced type in a field of science or learning customarily acquired by a prolonged course of specialized intellectual instruction and study." For example, does someone who completes a two-year, part-time training program in LAN administration meet the criteria to be classified as a professional?

From a practical standpoint, the IT industry recognizes people from a wide set of backgrounds, education, and personal experience in many different roles as IT professionals—provided they can do the job required by such a role.

Professional Relationships That Must Be Managed

IT professionals typically become involved in many different relationships: professional-employer, professional-client, professional-supplier, professional-professional, professional-IT user, and professional-society. In each relationship, an ethical IT professional will act honestly and appropriately and at all times steer away from any actions that would appear to be unethical. These various relationships are discussed below.

IT Professional-Employer

The IT professional-employer relationship is a critical one that requires ongoing effort by both parties to keep it strong. There are many facets to this relationship. The IT professional and employer discuss and agree upon fundamental aspects of this relationship before the professional accepts an employment offer. The aspects include job title, general performance expectations, specific work responsibilities, drug testing, dress code, location of employment, salary, work hours, and company benefits. Many IT professional-employer issues are addressed in the company's policy and procedures manual or in the company code of conduct, if it exists. These include protection of company secrets, vacation policy, time off for a funeral or illness in the family, tuition reimbursement, and use of company resources. Other aspects of the relationship come up over time as the need arises (for example, whether the employee can leave early one day if the time is made up on another day). Some aspects are addressed by law (for example, an employee may not be required to do anything illegal). Some aspects are specific to the role of the IT professional and are established based on the nature of the work or project—for example, the programming language to be used, the kinds and amounts of documentation to be produced, and the extent of testing to be conducted. Some of the more interesting ethical aspects of the employer-IT professional relationship are summarized in the following paragraphs.

As the stewards of an organization's IT resources, IT professionals must set the example and enforce the policies in regards to the ethical use of IT. IT professionals have the skills and knowledge to abuse systems and data or to allow others to do so. **Software piracy**, the act of illegally making copies of software or enabling others to access software to which they are not entitled, is an area where IT professionals can be tempted to violate laws and policies. Although end users get the blame, when it comes to using illegal copies of commercial software, software piracy in a corporate setting is sometimes directly traceable to IT people—either they are allowing it to happen or are actively engaging in it. Often the piracy is done to reduce IT-related spending to meet challenging budgets.

The cost of criminal or civil penalties to the corporation and the individuals involved, or even the cost of an out-of-court settlement with the offended software manufacturer, can easily be many times more expensive than the cost of "getting legal." The penalties can be up to $100,000 per copyrighted work if a software piracy case actually goes to trial and the defendant loses. In 2002, a Wisconsin printing company performed a self-audit and discovered that it had more copies of Adobe, Autodesk, Macromedia, Microsoft, and Symantec software programs on its computers than it had licenses to support. In addition to paying $150,000 to the Business Software Alliance (BSA), a software publishers' group, to avoid future legal problems, the company agreed to delete any unlicensed copies, purchase replacement software, and strengthen its software management practices.[2]

Trade secrecy is another area of the employer-professional relationship where issues often arise. A **trade secret** is a piece of information used in a business, generally unknown to the public, that the company has taken strong measures to keep confidential. It represents something of economic value, has required effort or cost to develop, and has some degree of uniqueness or novelty. Trade secrets can include a wide range of items, including the design of new software code, hardware designs, business plans, the design of the user interface to a computer program, and manufacturing processes. Specific examples include the Colonel's secret recipe of eleven herbs and spices, Wal-Mart's list of future store openings, and Intel's manufacturing process for producing the Pentium 4 chip. Employers fear that employees may reveal these secrets to competitors, especially when they leave the company. As a result, they have employees sign confidentiality agreements promising not to reveal the company's trade secrets. However, the IT industry is known for high employee turnover and things can get complicated when an employee moves on to a competitor. For example, in 2001, Compaq charged that a company founded by several former employees was using the computer giant's design and manufacturing trade secrets to develop competitive servers. Compaq further said that it believed that the company was engaging in "predatory targeting" of employees and customers of its server business. The company was served a restraining order stating that former Compaq employees cannot divulge trade secrets.[3]

Whistle-blowing is an effort by an employee of a company to attract the attention of others to a negligent, illegal, unethical, abusive, or dangerous act by the company that threatens the public interest. The whistle-blower often has special information about what is happening based on their expertise or position of employment within the offending organization. The goal of whistle-blowing is to fix a serious problem when working within the company to address the issue that has failed. For example, an employee within a chip manufacturing company may be aware that the manner in which a chemical used to make the chips is dangerous to employees and the general public. A conscientious employee would call the situation to management's attention and try to correct the problem by working with appropriate resources within the company. But what if the employee's attempt to correct the problem through internal channels is thwarted or ignored? The employee could then consider becoming a whistle-blower and point out the problem to people other than the employer, including state or federal agencies that have jurisdiction over the issue at hand. Obviously, such action on the part of the employee will have negative consequences on his or her career with the employer and could even result in retaliation against the employer and termination. Whistle-blowing is discussed more fully in Chapter 8.

IT Professional-Client

A client is someone for whom a professional provides services. A client may be someone from outside the professional's organization or it may be an "internal" client. In the IT professional-client relationship, each party agrees to provide something of value to the other. Generally speaking, the IT professional will provide a hardware or software product or services at a certain cost and within a given time frame. For example, the IT professional will agree to implement a new accounts payable system that meets defined user requirements for functionality on an agreed-to timetable. The client will provide compensation, access to key contacts, and perhaps work space. As a result, this relationship is often documented in contractual terms—who will do what, when the work will begin, how long it will take, how much the client will pay, and so on. Although there is typically a vast disparity in the knowledge or expertise of the parties in the IT professional-client relationship, the two parties must work together to be successful.

On a typical engagement, the client has the overall decision-making authority but makes decisions on the basis of information, alternatives, and recommendations provided by the IT professional. The client trusts the IT professional to use his or her expertise and to think and act in the best interests of the client. The IT professional must trust that the client will provide relevant information, listen to and understand what the professional says, ask questions to understand the impact of key decisions, and use that information to make wise choices between alternatives. Thus, the responsibility for decision-making is shared between client and professional. The following paragraphs discuss some of the ethical issues that can arise in the IT professional-client relationship.

During a project, there can be less than full and accurate reporting of project status by IT professionals. This may be because the IT project manager lacks the information, tools, or experience to perform an accurate assessment of project status. The project manager may wish to keep resources flowing into the project, hoping that the project problems can be corrected before anyone notices. The project manager may also be reluctant to share actual project status because of contractual penalties for failure to meet the schedule or to develop certain systems functions. In this situation, the client may not be informed about the problem until it has escalated to crisis level. After the full truth comes out, there can be finger pointing and heated discussions about cost overruns, missed schedules, and technical incompetence.

Another problem area of IT professional-client ethics is the tactic of IT consultants or auditors recommending their own products and/or services or those of an affiliated vendor to remedy a problem they have detected. For an example, an IT consulting firm is hired to assess a firm's IT strategic plan. After a few weeks of analysis, the consulting firm provides a poor rating for the existing strategy and insists that their proprietary products and services are required to develop a new strategic plan. Such a finding raises questions about the objectivity of the vendor and whether the vendor's recommendations can be trusted.

IT Professional-Supplier

IT professionals have dealings with many different hardware, software, and service providers. Most IT professionals understand that building a good working relationship with suppliers encourages the flow of useful communication and the sharing of ideas between the supplier and IT professional. Such information can lead to innovative and cost-effective ways of using the supplier's products and services that the IT professional may never have considered.

The IT professional must take steps to encourage such a relationship and to avoid doing things that would harm it. For example, the IT professional should be fair in dealing with suppliers and not make unreasonable demands. Saying "If you can't deliver the needed equipment tomorrow [say for an item where the normal industry lead time is one week], we will have to look at replacing you as our primary supplier" is aggressive behavior that does not encourage a good working relationship.

Suppliers also strive to maintain positive relationships with their customers in order to make and increase sales. Sometimes the actions taken to achieve this goal might be perceived as unethical—for example, by offering the IT professional a gift that is actually intended as a bribe. Clearly, the IT professional should not accept a bribe from a vendor, but one must be very careful in thinking about what constitutes a bribe. For example, acceptance of invitations to expensive dinners or payment of entry fees for a golf tournament may seem innocent to the recipient, but may be perceived as bribes by an internal accounting auditor.

IT Professional-Professional

Professionals feel a degree of loyalty to the other members of their profession. As a result, they are quick to help one another in obtaining new positions but slow to criticize one another in public. Professionals also have an interest in their profession as a whole, for how it is perceived affects how individual members are perceived and treated (for example, politicians, in general, are not thought to be very trustworthy, while teachers are). Hence, professionals owe one another an adherence to the profession's code of conduct. Experienced professionals can also serve as mentors and help develop new members of the profession. There are a number of ethical issues that can arise between members of a profession.

One of the IT profession's most common ethical problems is **resume inflation**, which involves lying on one's resume and saying that one is competent in an IT skill that is in high demand. Even though an IT professional might benefit in the short term from exaggerating his or her qualifications for a particular job or assignment, in the long run, such action hurts the profession, and in turn, individual practitioners. Customers, and society in general, become much more skeptical of IT professionals as a result.

Another ethical issue is the inappropriate sharing of corporate information. Due to the nature of their role, IT professionals have access to the corporate databases of private and confidential information about employees, customers, suppliers, new product plans, promotions, budgets, and so on. As discussed in Chapter 1, there have been many examples of such information being shared inappropriately. In some cases, such information is sold to other organizations. In other cases, such information is shared informally with others who have no need to know. This is done not in an attempt to gain financially, but simply in the course of conversation about work.

IT Professional-IT User

The term **IT user** is used to distinguish the person for whom a hardware or software product is designed from the IT professionals who develop, install, service, and support the product. The IT user employs the product to deliver organizational benefits or to increase his or her personal productivity.

IT professionals have a duty to understand the needs and capabilities of the user and to deliver the products and services that best meet those needs—subject, of course, to budget and time constraints. IT professionals also have a key responsibility to users to establish an environment that supports ethical behavior on the part of the user. Such an environment discourages software piracy, minimizes the inappropriate use of corporate computing resources, and avoids the inappropriate sharing of information. The establishment of an effective IT usage policy that addresses these issues will be covered later in this chapter.

IT Professional-Society

Regulatory laws establish safety standards for products and services to protect the public. However, these laws are less than perfect and fail to safeguard against all possible negative side effects of a product or process. Often the professionals doing the work can see more clearly the effect their work will have and can take action to eliminate potential public risks. Thus, society expects the members of a profession to practice their profession in ways that not only causes no harm to society, but also that provides significant benefits. One approach to accomplishing this is to establish and maintain professional standards that protect the public.

Clearly the actions of an IT professional can affect society. For example, a systems analyst may design a computer-based process control system that will be used to monitor a chemical manufacturing process. A failure or error in the system may put workers or residents in the neighborhood of the plant at risk. As a result, IT professionals have a relationship with others in society who may be affected by their actions. However, currently, there is presently no single, formal organization of IT professionals that is universally recognized as having responsibility to establish and maintain professional standards that protect the public.

THE ETHICAL BEHAVIOR OF IT PROFESSIONALS

Chapter 1 points out that the risks associated with inappropriate ethical behavior have grown in number, complexity, likelihood, and significance. As a result, corporations are taking a number of actions that apply to all employees to reduce business ethics risks. This section discusses actions to support the ethical behavior of IT professionals specifically.

Professional Codes of Ethics

A **professional code of ethics** states the principles and core values essential to the work of a particular occupational group. Practitioners in many professions subscribe to a code of ethics that governs their behavior. For example, doctors adhere to the 2000-year-old Hippocratic oath, varying versions of which medical schools offer as an affirmation to which their graduating classes assent. Most codes of ethics created by professional organizations have two main parts. The first outlines what the professional organization aspires to become, and the second typically lists rules and/or principles by which members of the organization are expected to abide. Many codes also include a commitment to continuing education as a fundamental tenet in recognition of the need for life-long learning by those who practice the profession.

Laws do not provide a complete guide to ethical behavior. Just because an activity is not defined as illegal does not mean that it is ethical. We also cannot expect a professional code of ethics to provide the complete answer—no code can be the definitive collection of behavioral standards. Only by understanding and adopting the principles behind the code and interpreting what is really intended can one achieve the intent. If one fully embraces and practices according to a professional code of ethics, there are many benefits—for the individual, for the profession, and for society as a whole. The following benefits of a code of ethics are discussed in the paragraphs that follow:

- Improves ethical decision-making
- Promotes high standards of practice and ethical behavior
- Enhances trust and respect from the general public
- Provides an evaluation benchmark

Improves Ethical Decision-Making

Adherence to a professional code of ethics means that practitioners will use a common set of core values and beliefs to serve as a guideline for ethical decision-making.

Promotes High Standards of Practice

Adherence to a professional code of ethics reminds professionals of the responsibilities and duties that they may be tempted to compromise to meet the press of day-to-day business. The code also defines the behaviors that are acceptable and unacceptable to guide professionals in their interactions with others. Strong codes of ethics actually have procedures for censuring professionals for serious violations, with penalties up to and including the loss of the right to continue to practice. Such codes are the exception, and no such codes exist in the IT arena.

Enhances Trust and Respect

Public trust is built on the expectation that a professional will behave ethically. We often find ourselves placed in a position where we must depend on the integrity and good judgment of a professional to tell the truth, abstain from giving self-serving advice, and offer warnings of the potential negative side effects of our actions. Thus, adherence to a code of ethics enhances our trust and respect of the professional and their profession.

Provides an Evaluation Benchmark

A code of ethics provides an evaluation benchmark that can be used by the professional as a means of self-assessment. Peers of the professional can also use the code for purposes of recognition or censure of individuals.

Professional Organizations

As of this time, no one IT professional organization has emerged as the preeminent organization, so there is no universal code of ethics for IT professionals. However, the existence of such organizations is useful in a field such as information technology that is rapidly growing and changing. The IT professional has a strong need to stay current about new developments in the field. This requires networking with others, seeking out new ideas, and building on one's current skills and expertise. Whether someone is a freelance programmer or the CIO of a Fortune 500 company, membership in an organization of IT professionals enables one to associate with others of similar work experience, to develop working relationships, and to exchange ideas. Information is disseminated from these organizations through e-mail, periodicals, Web sites, meetings, and conferences. Furthermore, in recognition of the need for professional standards of competency and conduct, many of these organizations have developed a code of ethics. Three of the most prominent IT professional organizations are summarized in the following paragraphs.

Association for Computing Machinery (ACM)

A computing society founded in 1947, the ACM serves more than 80,000 professionals in over 100 countries and offers more than 24 publications for technology professionals. *Tech News*, for example, is a comprehensive news-gathering service published three times a week. ACM's newest publication, *Ubiquity*, is a forum and opinion magazine. The organization also offers a substantial digital library of bibliographic information, citations, articles, and journals. ACM sponsors 37 special interest groups that focus on a wide variety of technology issues, including applied computing, data communications, and computer architecture. Each group provides publications, workshops, and conferences for information exchange.

The ACM has a code of ethics and professional conduct with supplemental explanations and guidelines. The ACM code consists of eight general moral imperatives, eight specific professional responsibilities, six organizational leadership imperatives, and two elements of compliance. The complete text of this code is provided in Appendix A.

Commitment to ethical professional conduct is expected of every member of the ACM. The code addresses many of the issues professionals are likely to face, but not all. The code is supplemented by a set of guidelines that provide explanation to assist members in dealing with the various issues contained in the code. The guidelines are also provided in Appendix A.

Association of Information Technology Professionals (AITP)

The AITP was formed in the 1950s to enable its members to keep ahead of the rapid pace of change in information technology. Its mission is to provide superior leadership and education in information technology. One of its goals is to help its members become more marketable in the broad and rapidly changing career field of information technology. The AITP publishes a monthly newsletter, *The Information Executive*, that focuses on current industry topics with contributions from industry experts, practitioners, and educators. The AITP also has a code of ethics and standards of conduct that are presented in Appendix B. The standards of conduct are considered to be rules that no true IT professional will violate.

Computer Society of the Institute of Electrical and Electronics Engineers (IEEE-CS)

The Institute of Electrical and Electronics Engineers (IEEE) covers the very broad fields of electrical, electronic, and information technologies and sciences. The IEEE-CS is one of the oldest and largest IT professional associations with over 100,000 members. Nearly a third of its members live and work outside the United States. Founded in 1946, it is the largest of the 36 societies of the IEEE. The IEEE-CS's vision is to be the leading provider of technical information and services to the world's computing professionals. The society promotes an active exchange of information, ideas, and technological innovation among its members through its many conferences, applications-related and research-oriented journals, local and student chapters, technical committees, and standards working groups.

In 1993, the IEEE-CS and the ACM formed a Joint Steering Committee for the Establishment of Software Engineering as a Profession. The initial recommendations of the committee were to define ethical standards, to define the required body of knowledge and recommended practices in software engineering, and to define appropriate curricula to acquire the body of knowledge. The purpose of the Task Force on Software Engineering Ethics and Professional Practices is to document the ethical and professional responsibilities and obligations of software engineers. Version 5.1 of the proposed draft of the *Code of Ethics for Software Engineers* can be found in Appendix C.[4]

The IEEE is concerned with a number of engineering disciplines other than computing. Thus, the IEEE also developed an earlier, shorter, and more general code of ethics that applies to a broad range of disciplines such as electrical, electronic, and information technologies and sciences. This code can be found in Appendix D.

Certification

Certification is a process that one undertakes voluntarily to prove competency in a set of skills. Vendors, training companies, and industry associations all grant various kinds of certifications. Because certification is no substitute for experience and doesn't guarantee that an individual will perform well on the job, hiring managers have become rather cynical on the subject. But IT certifications are still considered a useful way to measure knowledge, screen applicants, and assist in promotion decisions for entry-level IT positions or roles that require specific technical knowledge.

Vendor Certifications

Many IT vendors (Cisco, IBM, Microsoft, Oracle, and so on) offer certification programs for people who work with their products. Successful completion of such a certification program enables individuals to represent themselves as skilled in the use of the manufacturers' product(s). Certifications tied to a specific vendor's product are relevant for narrowly defined roles or certain aspects of broader roles. Sometimes, vendor certification is narrowly focused on technical details and does not address the strategic concepts that govern the entire subject area. For example, many IT professionals and trainers blamed insufficient discussion of customization to eliminate viruses in the Microsoft Certified Systems Engineer (MCSE) certification program for contributing to the spread of Code Red and other damaging viruses.[5]

The actual certification process requires passing a written exam (usually multiple choice because of legal concerns about objective grading of other types of exams). A few certifications (such as Cisco Systems' Cisco Certified Internetworking Engineer [CCIE]) require a hands-on lab exam to demonstrate skills and knowledge. It can take four or five years to obtain the necessary experience to pass some of the exams. Of course, there are courses and training material that can speed up the preparation process. Training costs can be expensive. Depending on the certification desired, the material needed for self-study courses can cost $1000 and in-class formal training courses can exceed $10,000.

So many people have obtained the (MCSE) certification (over 400,000 MCSEs have been certified in the United States) that this has become more of an entry-level requirement.[6] On the other hand, depending on the job market and the supply and demand for skilled workers, there are usually some certifications that can substantially improve one's salary or career prospects. Because of the rapid pace of change in the IT field, it is common for people to be recertified as newer technologies become available. For example, many people who were MCSE certified and trained on the Windows NT 4.0 operating system went through a recertification process for newer operating systems when they became available.

Industry Association Certifications

Certifications granted by industry associations generally ensure a certain level of experience and a broader perspective than vendor certifications; however, they often lag in developing tests covering new technologies. The trend in IT certification is toward a broadening from purely technical content to a mix of technical, business, and behavioral competencies—the competencies required in today's demanding IT roles. This is already evident in industry association certifications that address broader roles such as e-commerce, network security, and project management.

The Institute for Certification of Computing Professionals (ICCP) offers two levels of certification—Certified Associate Computing Professional and Certified Computing Professional. Since 1973, some 50,000 IT professionals worldwide have completed the ICCP certification process.[7] The ICCP certification procedures and exams are periodically reviewed and updated. Currently, candidates for either certificate must take a common core exam that includes questions on the topics of human and organization framework, systems concepts, data and information, systems development, technology, and associated disciplines.

The Associate Computing Professional (ACP) certification requires successful completion of an additional computer programming language exam. The ACP certification is for new entrants into the IT industry or for the recent college graduate who wishes to gain professional credentials substantiating his or her level of computing knowledge.

The Certified Computing Professional (CCP) certification is for the more experienced IT professional. It requires successful completion of exams on two of the following topics: management, procedural programming, business information systems, communications, office information systems, systems security, micro-computing and networks, systems development, software engineering, systems programming, and data resource management. In addition, the CCP candidate must have 48 months of full-time experience in information systems, or a bachelor's or master's degree in information systems or computer science with 24 months of full-time experience. All candidates must subscribe to a specified code of ethics, conduct, and good practice.[8]

The American Society for Quality Control (ASQC) has a certification process for a software quality engineer. The ASQC certification requirements include eight years of professional experience with at least three years in a decision-making position. A bachelor's degree may be counted as four years of experience, and an advanced degree may be counted as five years of experience. In addition, the individual must have proof of professionalism such as membership in an appropriate professional society, possession of a professional engineer's license, or statements from two professional colleagues. Finally, the individual must successfully complete a written exam in the areas of software quality management, software engineering, project management, appraisal, issues, analytical methods, and quality systems.

Clearly there are many different certifications available for the IT profession. The value of an IT certification for a given individual will vary greatly depending on where that person is in his or her career, what other certifications he or she possesses, and the nature of the IT job market.

Licensing

Licensing is a process generally administered at the state level in the United States. Certain professionals must undertake licensing to prove that they can practice their profession in a manner that is ethical and safe to the general public. Certified Public Accountants (CPAs), lawyers, doctors, various types of medical and daycare providers, and some engineers are examples of professionals that must be licensed by a state.

Various states have enacted legislation related to licensing. For example, the Texas Engineering Registration Act was enacted as the result of a tragic school explosion at New London, Texas, in 1937. It is the intent of this act that in order "to protect the public health, safety, and welfare that the privilege of practicing engineering be entrusted only to those persons duly licensed and practicing under the provisions of this Act." The act was updated and renamed the Texas Engineering Practice Act in the 1960s. Under the revised act, only duly licensed persons may legally perform engineering services for the public, and public works must be designed and constructed under the direct supervision of a licensed professional engineer. The terms "engineer" and "professional engineer" can be used only by persons who are currently licensed. Anyone who violates these parameters is subject to legal penalties.[9] Most states have similar laws. Thus, our definition of IT professional needs to take into consideration whether we are defining an IT professional in legal terms or in everyday terms.

The Case for Licensing IT Professionals

The days of simple, standalone information systems that operate completely independent from everything else are over. Modern information systems are highly complex, interconnected, and critically dependent on one another. Highly integrated enterprise resource planning systems provide information for decision-making to all the business functions (such as forecasting, production planning, purchasing, inventory control, manufacturing, and distribution) of the logistics supply chain of multi-billion-dollar corporations. Complex computer control and information systems manage and control the nuclear reactions of power plants used to generate electricity to run cities and nuclear submarines. Medical information systems monitor the vital statistics of hospital patients on critical life support. Local, state, and federal government information systems are entrusted with the responsibility of generating and distributing millions of checks worth billions of dollars to the public.

As a result of the increasing importance of information technology in our everyday lives, the development of reliable, effective information systems has become of mounting concern to the public. This has led many to consider whether the licensing of IT professionals would increase the reliability and effectiveness of information systems. Licensing would strongly encourage IT professionals to follow the highest standards of the profession and to practice a code of ethics. Licensing also allows for sanctions for violating best practices and the code of ethics. Without licensing, there are no requirements for heightened care and no concept of professional malpractice.

Issues Associated with Licensing IT Professionals

As discussed at the beginning of this chapter, the Texas Professional Engineers Licensing Board was working with the ACM to develop a licensing process for software engineering as a distinct discipline. This effort suffered a setback when the ACM announced that it is opposed to the licensing of software engineers.[10] Great Britain and the Canadian provinces of Ontario and British Columbia have adopted licensing for software engineers.[11] A professional engineer exam for "Computer Engineer" is under development by the National Council of Engineering Examiners and Surveyors (NCEES). However, there is currently no international or national licensing program for IT professionals, and there are many reasons for this:

- **There is no universally accepted core body of knowledge.** The core body of knowledge (CBOK) for any profession outlines agreed-upon core areas of knowledge, skills, and abilities that all licensed professionals must possess. At present, there is no universally accepted CBOK for

IT professionals on which to base licensing. Instead, several different professional societies have developed their own CBOK for different IT professional roles such as programmer and software engineer. In addition, some state legislatures are attempting to define a CBOK to regulate the practice of software engineering in their states.

- **It is unclear who should manage the content and administration of licensing exams.** How will licensing exams be constructed and who will be responsible for designing and administering the exams? Will someone who passes a license exam in one state or country be accepted as licensed in another state or country? In a field as rapidly changing as IT, it is clear that the professional must make a commitment to ongoing, continuous education. Should the IT professional license expire every few years (like a driver's license), at which time the practitioner must prove competence in new practices related to his or her area of specialization to renew the license? These questions would normally be answered by the state licensing agency responsible for licensing other professionals.

- **There is no administrative body to do accreditation of professional education programs.** Unlike the American Medical Association (AMA) for medical schools, or the American Bar Association (ABA) for law schools, there is no single body that performs accreditation of IT professional education programs. Furthermore, there is no well-defined, step-by-step process to train someone for a role as an IT professional—even if one focuses on a very specific role, such as that of programmer. Instead, there are many different approaches. Indeed, there is not even broad agreement on what skills one must possess to be an excellent programmer—it is highly situational, depending on the computing environment in which one is working.

- **There is no administrative body to assess and assure competence of individual professionals.** As lawyers, doctors, and other licensed professionals conduct their practice, they are held accountable to high ethical standards and can lose their license for failing to meet these standards or for demonstrating incompetence. AITP standards of conduct state: "Take appropriate action in regard to any illegal or unethical practices that come to my attention. However, I will bring charges against any person only when I have reasonable basis for believing in the truth of the allegations and without any regard to personal interest." When one looks at the other IT professional codes of ethics, there is nothing that addresses the censure issue quite so strongly as the AITP code. However, the truth of the matter is that the AITP standards of conduct have seldom, if ever, been used to censure practicing IT professionals.

Some people believe that the licensing of IT professionals will increase the likelihood of an IT professional being sued for negligence or professional malpractice. Read the following Legal Overview and see what you think.

LEGAL OVERVIEW

PROFESSIONAL MALPRACTICE

Negligence is defined as "the omission to do something which a reasonable man would do, or doing something which a prudent and reasonable man would not do."[12] **Duty of care** refers to the obligation that we not cause any unreasonable harm or risk of harm to others. For example, each person owes a duty to keep one's pets from attacking others and to operate one's car in a safe manner. Similarly, businesses owe a duty to not discharge dangerous pollutants into the air or water, to make safe products, and to not cause employees to suffer accidents due to unsafe operating conditions.

The courts decide whether a duty of care is owed in specific cases by applying a **reasonable person standard**. The defendant's actions are evaluated against how an objective, careful, and conscientious person would have acted in the same circumstances. Defendants with a particular expertise or competence are measured against a **reasonable professional standard**. For example, consider a medical malpractice suit involving personal injuries arising from improper treatment of a broken bone. The reasonable person standard would be set higher if the plaintiff were an orthopedic surgeon rather than a general practitioner. In the IT arena, consider a negligence case in which the actions of an employee resulted in the permanent loss of millions of customer records. The reasonable person standard would be set higher if the plaintiff were a licensed IT professional who is also an Oracle-certified database administrator (DBA) with 10 years of experience versus an unlicensed systems analyst with no DBA experience and no specific knowledge of the Oracle database management system.

Should the court find that the defendant actually owed the plaintiff a duty of care, it must determine whether the defendant breached this duty. A **breach of the duty of care** is the failure to act as a reasonable person would act. A breach of duty may consist of either an action (such as throwing a lit cigarette into a fireworks factory and causing an explosion) or a failure to act when there is a duty to act (such as a police officer failing to protect a citizen from an attacker).

A professional who breaches this duty of care is liable for the injury his or her negligence causes. This liability is commonly referred to as **professional malpractice**. For example, a certified public accountant who fails to use reasonable care, knowledge, skill, and judgment when auditing a client's books is liable for accounting malpractice. Professionals who breach this duty are liable to their patients or clients. They may also be liable to some third parties.

Courts have consistently rejected attempts to sue individuals for computer-related malpractice. Professional negligence requires that harm be caused by the failure of a person to perform within the standards of his/her profession. To date, software engineering is not a uniformly licensed profession in the United States and there are no uniform standards against which to compare one's professional behavior. Thus, software engineers cannot be subject to malpractice lawsuits.

IT USERS

Chapter 1 outlines a number of actions that corporations are taking to reduce the increasing risk associated with unethical behavior. These actions applied to all employees and were not specific to IT issues. The next section discusses actions specific to improving the ethical behavior of employees in regard to their use of IT. This is an area of rapidly growing concern as more and more companies provide their employees with personal computers and with the corresponding access not only to corporate information systems and data, but also to the Internet and the World Wide Web (WWW).

IT Users and Common Ethical Issues

A few of the more common IT user ethical issues will be mentioned in the following paragraphs. Other ethical issues will be discussed in future chapters.

Software Piracy

As mentioned, software piracy in a corporate setting is sometimes directly traceable to IT professionals—either they are allowing it to happen or they are actively engaging in it. IT users should be encouraged by corporate IT usage policies and management to report instances of piracy and to challenge why the practice is being allowed.

There are also times when IT users are the perpetrators of the act of software piracy. A common violation occurs when IT users copy software from their work computers to take home to load onto their home computer. When confronted, the IT user's argument might be this: "I bought a home computer partly so I could take work home and be more productive; therefore, I need the same software on my home computer as I have at work." However, this is still piracy because neither the employee nor the employer has paid for an additional license for the software used on the employee's home computer.

Inappropriate Use of Computing Resources

It is not unusual for employees to explore the capabilities of their IT resources. For some, this means using their workstation to surf popular, nonbusiness Web sites, participate in chat rooms, view pornographic sites, and play computer games. Clearly, these are not activities that were envisioned by management when it provided valuable IT resources. These non-work-related activities eat away at worker productivity by taking up the employees' time at work. Furthermore, viewing sexually explicit material, sharing lewd jokes, and sending hate e-mail is considered unacceptable and could involve the employer in a harassment suit alleging that the company failed to control its employees and allowed a work environment conducive to racial or sexual harassment. For example, an allegedly racist e-mail message was at the center of a $30 million discrimination suit filed in 1998 against Morgan Stanley, Dean Witter & Co.

Inappropriate Sharing of Information

Every organization stores vast amounts of information that can be classified as either private data or confidential information. Private data is data about individual employees—for example, salary information, time and attendance data, health records, performance ratings, and other personal background information. Confidential information is data about the company and its operations—sales and promotion plans, staffing projections, manufacturing processes, product formulae, tactical and strategic plans, and research and development results. When an IT user either inadvertently or intentionally shares this information with someone unauthorized to have it, someone's personal privacy has been violated or the potential has been created for key company information to fall into the hands of competitors. For example, if an IT employee with access to payroll records were to discuss the salary of a coworker with a friend, this would be a clear violation of the coworker's privacy.

Supporting the Ethical Practices of IT Users

Obviously, the introduction of information technology has raised the potential for new ethical issues and problems. Many organizations have recognized the need for a policy on the use of information technology to protect against the abuses previously discussed. Although no policy can stop someone from doing something wrong, such a policy can set forth the general rights and responsibilities common to all users of information technology, establish the boundaries of acceptable and unacceptable behavior, and enable management to take action against those who violate the policy. If there is adherence to it, such a policy also has the added benefits of improving services to users, increasing productivity, and reducing costs to the organization. Companies can take several actions when creating an IT usage policy.

Define and Limit the Appropriate Use of IT Resources

It is imperative that each company develop, communicate, and enforce written guidelines regarding the appropriate use of IT resources. Such guidelines encourage employees to respect the company's IT resources and use them to enhance their job performance. Workable yet effective guidelines are those that allow for some level of personal use while requiring that employees not visit objectionable Internet sites nor use company e-mail assets to send objectionable or harassing information.

Establish Guidelines for Use of Company Software

The IT managers in an organization must provide clear rules governing the use of home computers and associated software. Of course, some companies provide both personal computers and software for IT users to be productive at home. Other companies make it possible for employees to buy the hardware and/or software at corporate discount rates. Still others negotiate contracts with software manufacturers that allow for the use of corporate software on the home computers of employees. The goal should be to ensure that individuals have legal copies of all the software they need to be effective in their roles while they are at work, on the road, or at home—if they are expected or encouraged to work at home.

Structure Information Systems to Protect Data and Information

It is crucial that organizations implement systems and procedures that limit the access to data to those employees who have a "need to know." For example, sales managers may have complete and total access to sales and promotion plan databases through the company network, but their access to specific data should be limited to information only about those products for which they are responsible. Furthermore, they should be prohibited from accessing data about research and development results, product formulae, and staffing projections. Their access to these other kinds of data is prohibited because they do not need this information to perform their role effectively.

Install and Maintain a Corporate Firewall

An **firewall** is a hardware and/or software device that serves as a barrier between a company and the outside world and limits access into and out of the company's network based on the organization's Internet usage policy. The firewall can be configured to serve as an effective deterrent to non-work-related Web surfing by blocking access to specific, objectionable Web sites. Unfortunately, the number of such sites grows so rapidly that it is difficult to specify the URLs of all such sites to the firewall. The firewall can also serve as an effective barrier to incoming email from certain Web sites, companies, or individuals. It can even be programmed to not allow email with certain kinds of attachments (for example, Word documents) to pass through. This reduces the risk of harmful computer viruses.

Table 2-1 presents a manager's checklist that summarizes the items to consider when establishing an IT usage policy. The preferred answer in each case is "Yes."

Table 2-1   Manager's checklist of items to consider when establishing an IT usage policy

Questions	Yes	No
Is there a statement making clear the need for an IT usage policy?	___	___
Does the policy provide a clear set of guiding principles that can be used for ethical decision-making?	___	___
Is it clear how the policy applies to the following types of workers?	___	___
Employees	___	___
Part-time workers	___	___
Temps	___	___
Contractors	___	___
Does the policy address the following issues:	___	___
Protection of the data privacy rights of employees, customers, suppliers, and others?	___	___
Limits and control of access to proprietary company data and information?	___	___
The use of unauthorized or pirated software?	___	___
Employee monitoring, including e-mail, wiretapping and/or eavesdropping on phone conversations, computer monitoring, and surveillance by video?	___	___
Respect of the intellectual rights of others including trade secrets, copyright, patents, and trademarks?	___	___
Inappropriate use of IT resources such as nonbusiness related Web surfing, sending and receiving of nonbusiness related e-mail, and use of computers for other than business?	___	___
The need to protect the security of IT resources through adherence to good security practices such as not sharing user IDs and passwords, use of "hard-to-guess" passwords, and frequent changing of passwords?	___	___
The use of the computer to intimidate, harass, or insult others through the use of abusive language in e-mails and other means?	___	___
Are disciplinary actions for IT-related abuses defined?	___	___
Is there a process for communicating the policy to employees?	___	___
Is there a plan to provide effective, on-going training relative to the policy?	___	___
Has a corporate firewall been implemented?	___	___
Is the corporate firewall maintained?	___	___

Summary

1. What are the key characteristics that distinguish a professional from other kinds of workers and what is the role of an IT professional?

   A professional is someone who requires advanced training and experience, must exercise discretion and judgment in the course of his or her work, and whose work cannot be standardized. A professional is expected to contribute to society, to participate in a life-long training program (both formal and informal) to keep abreast of developments in his or her field, and to help develop other professionals. In a legal sense, a professional is one who has passed the state licensing requirements (if they exist) and earned the right to practice as a professional in that state. The IT industry recognizes many people in many different roles as IT professionals—provided they can do the jobs required by their roles.

2. What are the various professional relationships that must be managed by the IT professional and what are the key ethical issues that can arise in each?

   IT professionals typically become involved in many different relationships and each type of relationship has its own set of ethical issues and potential problems. For the IT professional-employer relationship, some of the more important ethical issues involve setting and enforcing policies in regard to the ethical use of IT, the potential for whistle blowing, and the safeguarding of trade secrets. For the IT professional-client relationship, the key issues revolve around defining, sharing, and fulfilling the responsibilities of each party in the successful completion of a consulting engagement. For the IT professional-supplier relationship, a major ethical issue is working to develop a good working relationship in such a way as to do nothing that can be perceived as acting in an unethical manner. For the IT professional-professional relationship, the key issues are working to improve the profession through such activities as mentoring inexperienced members of the profession and demonstrating professional loyalty. Resume inflation and the inappropriate sharing of corporate information are also relevant issues. For the IT professional-IT user relationship, some important issues include software piracy, inappropriate use of IT resources, and inappropriate sharing of information. For the IT professional-society relationship, the main issue is for IT professionals to practice their profession in ways that not only cause no harm to society, but also that provide significant benefits.

3. How do codes of ethics, professional organizations, certification, and licensing affect the ethical behavior of IT professionals?

   A professional code of ethics states the principles and core values essential to the work of a particular occupational group. Such a code serves as a guideline for ethical decision-making, promotes high standards of practice and ethical behavior, enhances trust and respect from the general public, and provides an evaluation benchmark.

   Many people believe that the licensing and/or certification of IT professionals would increase the reliability and effectiveness of information systems. There are many issues associated with the licensing of IT professionals, including a) there is no universally accepted core body of knowledge on which to test people for licensing, b) it is unclear who should manage the content and administration of licensing exams, c) there is no administrative body to do accreditation of professional education programs, and d) there is no administrative body to assess and assure competence of individual professionals.

4. What are the key tenets of four different codes of ethics that provide guidance for IT professionals?

   Several IT professional organizations have developed a code of ethics, including the ACM, the AITP, and the IEEE-CS. These codes have two main parts—the first outlines what the organization aspires to become and the second typically lists rules and/or principles that members of the organization are expected to live by. They also include a commitment to continuing education as a fundamental tenet in recognition of the need for life-long learning by those who practice the profession.

5. What are the common ethical issues that face IT users?

IT users often encounter ethical issues related to the use of IT, including software piracy, inappropriate use of corporate IT resources, and the inappropriate sharing of private and secret information.

6. What approaches can be taken to support the ethical practices of IT users?

The development of an IT usage policy is the first step for an organization in defining appropriate and inappropriate IT user behavior. The policy should define and limit the appropriate use of IT resources and set forth clear guidelines for use of company software. In addition, IT professionals within the organization can also structure information systems and establish corporate firewalls to support IT users' appropriate use of IT resources.

Review Questions

1. How do you distinguish a professional from other kinds of workers? How would you define the term *IT professional*?
2. What are the six relationships in which an IT professional becomes involved? Identify at least one key issue for each of these relationships.
3. What is whistle-blowing?
4. How do you define a trade secret? What actions might an organization take to protect its trade secrets?
5. What is software piracy? How might IT professionals be involved in software piracy?
6. What is a professional code of ethics? How is it different from the corporate code of conduct discussed in Chapter 1?
7. List three benefits associated with adherence to a code of professional ethics.

8. Identify the three most prominent IT professional organizations. What are some benefits of membership in each of these organizations?
9. What is negligence? What must a plaintiff show to prove negligence?
10. What is the difference between licensing and certification?
11. Identify three benefits of licensing IT professionals.
12. Identify and discuss four issues associated with the licensing of IT professionals.
13. What are some common ethical issues encountered by IT users? What negative impact does unethical behavior in these areas have?
14. What are three actions that can be taken to strengthen the ethical practices of IT users?

Discussion Questions

1. What is wrong with resume inflating if it enables one to obtain a better job?
2. What are some of the benefits of joining a professional organization? What might be some disadvantages?
3. What points can you present in favor of licensing of IT professionals? What points can you present that argue against licensing?

4. Review the ACM *Code of Ethics* provided in Appendix A. The code covers many of the issues an IT professional is likely to face, but not all. Identify two key issues not addressed by the ACM code of ethics.
5. Most codes take a strong position on policing incompetence in the profession. The AITP standards of conduct state the following: "Take

appropriate action in regard to any illegal or unethical practices that come to my attention. However, I will bring charges against any person only when I have reasonable basis for believing in the truth of the allegations and without any regard to personal interest." There is nothing that addresses a similar issue quite so strongly in the ACM, IEEE, or software engineering codes. Why do you think these other codes do not mention the issue of policing incompetence?

6. Should all IT professionals either be licensed or certified? Why or why not?

7. What commonalities do you find among the IT professional codes of ethics discussed in this chapter? What differences are there? Are there any issues important to you that are not addressed by these codes of ethics?

What Would You Do?

1. In Italy, *raccomandazione* is the custom of seeking and receiving special treatment from people in power or from people who are close to someone in power. The ability to solicit favors from someone in a higher place, be it through the chief of police or the chief of police's chauffeur, has been part of the Italian art of getting things done for over 2000 years. In April 2001, Italy's highest court of appeal ruled that influence peddling is not a crime. The judges did rule, however, that it is a crime to overstate one's power in order to exert influence.

   Your firm is opening a new sales office in Rome and will be using a local employment agency to identify and screen candidates who will then undergo employment testing and interviews by members of your organization. What guidelines would you provide to the agency in regard to the practice of raccomandazione to ensure that the employment agency operates ethically and effectively?

2. Kim is more than your boss. She's the one who recommended you for your fast-track new job. You regularly have lunch with her and you even play tennis together. Over lunch one day, while talking about the pending desktop computer upgrade, she lets it slip that ABCXYZ Computer donated $5,000 to "my favorite charity." Two weeks later, you learn that ABCXYZ has won the contract,

even though your company has had reliability problems with ABCXYZ's products in the past. What would you do?

3. Jacob is the vice president of sales and an important ally of your IT department. He's gone to bat for you before the CEO on important IT projects, such as a big sales automation system, and has valuably assisted in advocating for the use of the latest software packages within the sales organization. Jacob has played a major role in your success so far. You've just learned that Jacob and his support staff have acquired and are using an unlicensed Lotus software suite on their desktops, while the rest of the company is standardized on Microsoft's Office Suite. You've talked to him and the rest of the company's leadership team about the need for standardized software and the risks the company runs if it uses unlicensed software, but no action has been taken. What would you do?

4. You are the new CIO at a small manufacturing company with 500 employees located at one plant, two warehouses, and a headquarters building. Your manager is the chief financial officer (CFO) and she has asked that you make it a high priority to establish a set of policies and guidelines on the use of IT resources—the firm currently has none. How would you proceed with this?

1. Online Brokers Experience Problems

In January 2001, The Office of Compliance Inspections and Examinations of the Securities and Exchange Commission (SEC) released a report calling for brokerage firms and securities dealers to evaluate their online trading programs—especially their processing capabilities. The most common complaints submitted to the SEC by users of Web trading sites are related to inadequate operational capabilities—failures or delays in processing orders online, difficulty in accessing online accounts, and errors in processing orders. As a preventive measure, the SEC recommended that online brokers maintain detailed records of capacity evaluations, system slowdowns, and systems outages, including information about the causes and impacts of problems. The SEC also urged companies to use every reasonable effort to notify customers when they are experiencing operational difficulties.

On February 12, 2001, Charles Schwab & Co. was forced to switch its online trading Web site to a backup server for six hours. During this time period, the discount broker's customers were unable to either receive trade confirmation messages or view previous transaction records. As a backup process, Schwab customers were notified to contact Schwab representatives by telephone. Customer database problems were caused by a glitch in off-the-shelf software or a homegrown system, not processing capacity limitations. This disruption occurred barely one month after the SEC issued its warning to brokerage firms and securities dealers; however, no SEC fines or sanctions were levied against Charles Schwab as a result of this incident.

In the same month as the Charles Schwab incident, the New York Stock Exchange (NYSE) fined online brokerage TD Waterhouse Investor Services $225,000 and censured it for problems related to Web site failures that temporarily stopped it from filling online stock orders and for inadequate customer service related to the outages. According to a NYSE statement, TD Waterhouse was unable to process online customer orders on 33 different trade days over an 18-month period starting in late 1998. During this time period, TD Waterhouse had approximately 250,000 online customer accounts placing an average of 48,000 trades per day with webBroker, its online order entry system. The Web site failures ranged from a couple of minutes to nearly two hours.

The NYSE also said in its decision that TD Waterhouse didn't maintain adequate telephone routing systems to handle orders that would have been placed online, resulting in lengthy telephone hold times for customers. Furthermore, TD Waterhouse failed to adequately advise customers of an alternative touch-tone telephone order entry system, TradeDirect, that was available during all webBroker outages. As a result, many customers were unable to place orders.

To further compound their problems, it is alleged that TD Waterhouse failed to report some 18,000 verbal and 2300 e-mail complaints related to these outages to the SEC as required by Exchange Information Memorandum 98-3. The statistical information on customer complaints is important because the information discloses trends and issues that may be used in determining the focus of future SEC examinations.

TD Waterhouse accepted the penalties without admitting or denying its guilt. A TD Waterhouse spokesperson said that the firm was sanctioned for the way it handled the outages, not the actual outages. The software issues that caused the outages have been corrected and no webBroker outages have been reported since April 2001.

Questions:

1. What material differences exist between the outage experienced by Charles Schwab versus those of TD Waterhouse? Do you think that the differences were significant enough that it was fair that the one firm received no fine or sanctions while the other was heavily fined? Why or why not?

2. Do any elements of negligence present themselves in the actions of Charles Schwab? TD Waterhouse? Which company might be at greater risk of a negligence lawsuit on behalf of its customers? Why?

3. Imagine that you are an outside consultant assigned to a TD Waterhouse team charged with responsibility to "put an end" to customer service interruptions. One focus area you wish to explore is the need for a professional code of ethics, licensing, and/or certification of employees that have anything to do with the online brokerage systems. How would you assess if such actions are appropriate?

Sources: adapted from Lucas Mearian, "NYSE Fines TD Waterhouse for Web Site Failures," *Computer World*, March 1, 2001 accessed at www.computerworld.com/; Craig Stedman, "SEC Staff: Online Brokerages Can Do Better," *Computer World*, January 31, 2001, accessed at www.computerworld.com/; Maria Trombly, "Schwab Database Glitch Cuts Users Off from Some Information," *Computer World*, February 13, 2001, accessed at www.computerworld.com; and About Us – Company Profile Section of TD Bank Financial Group Web site accessed at www.td.com/ on May 25, 2001.

2. IT Usage Policy

Read the proposed policy on the use of IT technology for the University of Cincinnati (it's printed after the list of questions) and use the manager's checklist to answer the following questions:

1. Are all of the key issues covered by this policy? If not, which ones need to be addressed?
2. Is the statement of enforcement clear and strong? If not, how would you reword this section of the policy?
3. How would you ensure that this policy is communicated and understood by the broad group of IT users at the university—students, professors, research people, administrative support staff, contractors, and part-time workers?
4. Examine the IT usage policy in effect at your school. Write a paragraph identifying its strengths and weaknesses.

General Policy on the Use of Information Technology
Proposed September 2000
Introduction
As an institution of higher learning, the University both uses information technology and supplies it to the members of the university community. This policy sets forth the general rights and responsibilities common to all uses of information technology, from the simple stand-alone PC to the complex systems that create virtual classrooms, workplaces and recreational facilities in the University.

This policy applies to all members of the University community, including guests who have been given accounts on the University's information technology systems for specific purposes. It also applies whether access is from the physical campus or from remote locations. In addition, there may be specific policies issued for individual systems, departments, colleges and the like. While these policies must be consistent with this general policy, they provide more detailed guidance about what is allowed and what is prohibited on each system. All members of the University community are responsible for familiarizing themselves with any applicable policy prior to use.

Guiding Principles
The primary guiding principle is that the rules are the same for information technology as for other aspects of university life. The rights and responsibilities governing the behavior of members of the University community are the same on both the virtual and physical campuses, and the same disciplinary procedures will be followed when the rules are violated. There is nothing special about the virtual campus that makes it distinctly different.

The University has a strong commitment to the principles of free speech, open access to knowledge, and respect for a diversity of opinions. The rights as well as the restrictions governing these principles on the physical campus apply fully to the virtual campus.

Specific Areas

1. Applicable Laws and Regulations

All members of the University community must obey:

- All relevant federal, state, and local laws. These include laws of general application such as libel, copyright, trademark, privacy, obscenity, and child pornography laws as well as laws that are specific to computers and communication systems, such as the Computer Fraud and Abuse Act and the Electronic Communications Privacy Act.
- All relevant University rules and regulations. These include the Rules of the University, the Student Code of Conduct, the various collective bargaining agreements between the University and its employees, and all other University policies including the policy against sexual and racial harassment.
- All contracts and licenses applicable to the resources made available to users of information technology.
- This policy as well as other policies issued for specific systems.

2. Resource Limits

Information technology resources are often limited; what is used by one person is no longer available to others. Many systems have specific limits on several kinds of resources, such as storage space or connect time. All users must comply with these limits and not attempt to circumvent them. Moreover, users are expected not to be wasteful of resources whether or not there are specific limits placed on them. Unreasonable use of resources may be curtailed.

3. Privacy

Members of the University community shall not attempt to access the private files of others. The ability to access a file does not, by itself, constitute authorization to do so.

The University does not routinely monitor or inspect individual accounts, files, or communications. There are situations, however, in which the University has a legitimate need to do so: (1) system managers may access user accounts, files, or communications when there is reason to believe that the user is interfering with the performance of a system; (2) authorized investigators may access accounts, files, or communications to obtain relevant information when there is a reasonable suspicion that the user has violated either laws or University policies; (3) co-workers and supervisors may need to access accounts, files, or communications used for university business when an employee becomes unavailable; and (4) when required by law. All monitoring and inspection shall be subject to authorization, notification and other requirements (forthcoming).

Though the University will attempt to prevent unauthorized access to private files, it cannot make any guarantees. Because the University is a public entity, information in an electronic form may be subject to disclosure under the Ohio Public Records Act just as paper records are. Information also can be revealed by malfunctions of computer systems, by malicious actions of hackers, and by deliberate publication by individuals with legitimate access to the information. Users are urged to use caution in the storage of any sensitive information.

4. Access

Some portions of the virtual campus, such as public web pages, are open to everyone. Other portions are restricted in access to specific groups of people. No one is permitted to enter restricted areas without authorization or to allow others to access areas for which they are not authorized. The ability to access a restricted area does not, by itself, constitute authorization to do so.

Individual accounts are for the use of the individual only; no one may share individual accounts with anyone else, including members of the account holder's family. Joint access to resources when needed, should be provided from separate accounts.

5. Security

All members of the University community must assist in maintaining the security of information technology resources. This includes physical security, protecting information and preventing and detecting security breaches. Passwords are the keys to the virtual campus and all users are responsible for the security of their passwords. Users must report all attempts to breach the security of computer systems or networks to an appropriate official.

6. Plagiarism and Copyright

Intellectual honesty is of vital importance in an academic community. You must not represent the work of others as your own. You must respect the intellectual rights of others and not violate their copyright or trademark rights. It is especially important that you obey the restrictions on using software or library resources for which the University has obtained restricted licenses to make them available to members of the University community.

7. Enforcement

Anyone who becomes aware of a possible violation of this policy or the more specific regulations of the systems that comprise the virtual campus should notify the relevant department head or system administrator. The administrator will investigate the incident and determine whether further action is warranted. The administrator may resolve minor issues by obtaining the agreement that the inappropriate action will not be repeated. In those cases that warrant disciplinary action, the system administrator will refer the matter to the appropriate authorities. These include Public Safety for violations of criminal law, the Office of Student Affairs for violations by students, the appropriate Provost for violations by faculty, and the Office of Human Resources for violations by staff members.

System administrators can act to block access and disable accounts when necessary to protect the system or prevent prohibited activities, but such actions cannot be used as punishments. Users must be notified promptly of the action and the restrictions must be removed unless the case is referred for disciplinary action.

Source: University of Cincinnati Information Technologies Office.

Endnotes

1 "ACM Position Papers on Software Engineering & Licensing," American Computing Machinery Web site at www.acm.org, accessed on May 24, 2002.

2 Press Room, "Wisconsin Company Pays Software Watchdog $150,000," April 29, 2002, accessed at the Business Software Alliance Web site at www.bsa.org.

3 James Evans, "Compaq Sues Server Start-up Over Trade Secrets," *ComputerWorld*, March 1, 2001, accessed at www.computerworld.com on April 25, 2001.

4 IEEE Computer Society Web site at www.computer.org accessed on April 12, 2001.

5 Dan Verton, "Microsoft MCSE Training Faulted," *Computerworld*, August 13, 2001, accessed at www.computerworld.com.

6 Mary Brandel, "The Top Certifications," *Computerworld*, May 14, 2002, accessed at www.computerworld.com.

7 The ICCP Web site accessed at www.iccp.org on April 27, 2001.

8 The ICCP Web site accessed at www.iccp.org on April 27, 2001.

9 "Licensing Information—Who Should Be Licensed," Texas Board of Professional Engineers Web site at www.tbpe.state.tx.us accessed on May 25, 2002.

10 "ACM Council Resolutions on Software Engineering Licensing," May 16, 1999, accessed at the ACM Web site www.acm.org.

11 Nancy Mead, "Issues in Licensing and Certification of Software Engineers," accessed at the Carnegie Mellon Software Engineering Institute Web site at www.sei.cmu.edu.

12 Henry R. Cheeseman, "Contemporary Business Law," 3rd edition, @ 2000, Prentice-Hall: Upper Saddle River, NJ, pg. 88.

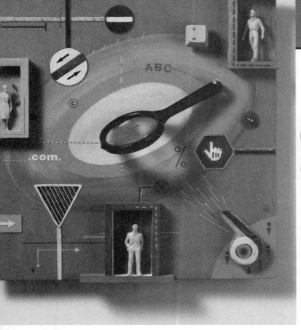

Chapter 3

Computer and Internet CRIME

"U.S. businesses will increasingly become the point of attack for enemies of [the] United States by hackers and national governments using sophisticated weapons such as worms and viruses that can be used for precise attacks." [1]
— Lawrence Gershwin, national intelligence officer at the CIA

In April 2002, computer hackers calling themselves the "Deceptive Duo" announced that they had begun their mission of breaking into computer systems to call attention to vulnerabilities in the United States' national security. The hackers said they believe that breaking into computer systems is the only way to get some organizations to take action to improve security.

In subsequent weeks, they hacked into over 52 Web sites and databases, including those operated by the United States Office of the Secretary of Defense, the Space and Naval Warfare Systems Command, the Defense Logistics Agency, Sandia National Laboratories (which performs a wide variety of national security research and development work), the NASA Jet Propulsion Laboratory (the lead center for unmanned space exploration), Midwest Express Airlines, and a number of banks. After information was acquired, they targeted an appropriate Web site to post screenshots of the data. For example, data from the Defense Logistics Agency database was posted on a Web site of the Office of the Secretary of Defense.

A spokesperson for the Space and Naval Warfare Systems Command admitted that hackers had gained access to their system because the agency had failed to change the default password and administrator's user ID—an obvious mistake on the part of the agency.

Reaction to the Deceptive Duo has been strongly divided. Some believe that they have an ulterior motive—to gain publicity, to promote computer security products, or to add excitement to their otherwise boring lives. Many believe that their actions are harmful and illegal, comparing them to robbers who break into a home and leave a note saying the locks need to be changed. Some defend the Deceptive Duo and say they are doing the public a service by bringing attention to computer security deficiencies.†

Sources: adapted from Linda Rosencrance, "Hacker Duo Says They Hack for the Sake of National Security," *Computerworld*, May 2, 2002, accessed at www.computerworld.com; Linda Rosencrance, "Update: Feds Seize Equipment from 'Deceptive Duo' Suspect," *Computerworld*, May 17, 2002, accessed at www.computerworld.com; Dennis Fisher, "'Deceptive Duo' Strikes Again," *eWeek*, May 3, 2002, accessed at www.eweek.com; and Dennis Fisher, "Feds Raid Suspected 'Deceptive Duo' Digs," *eWeek*, May 15, 2002, accessed at www.eweek.com.

As you read this chapter, consider the following questions:

1. What are some key trade-offs and ethical issues associated with the safeguarding of data and information systems?

2. Why has there been a dramatic increase in the number of computer-related security incidents in recent years?

3. What are the most common types of computer security attacks?

4. What are some of the characteristics of common computer crime perpetrators, including their objectives, available resources, willingness to accept risk, and frequency of attack?

5. What are the key elements of a multi-level process for managing security vulnerabilities based on the concept of reasonable assurance?

6. What actions must be taken in response to a security incident?

IT SECURITY INCIDENTS: A WORSENING PROBLEM

The security of information technology used in business is of utmost importance. Confidential business information and private customer and employee information must be safeguarded, and systems must be protected against malicious acts intended to steal data or otherwise cause disruption. Although the necessity of security is obvious, this need often must be balanced against other business needs and issues. Business managers, IT professionals, and IT users all face a number of ethical decisions regarding IT security, including the following:

- If my firm is a victim of a computer crime, should we pursue prosecution of the criminals at all costs or maintain a low profile to avoid the negative publicity?
- How much effort and money should be spent implementing safeguards against computer crime (how safe is safe enough)?
- If my firm produces software that contains defects that allow hackers to attack my customers' data and computers, what actions should we take?
- What tactics should management request that employees use to gather competitive intelligence without doing anything illegal?
- What should be done if recommended computer security safeguards make life more difficult for customers and/or employees, resulting in lost sales and increased costs?

Unfortunately, there is no doubt that the number of IT-related security incidents is increasing—not only in the United States, but around the world. The Computer Emergency Response Team Coordination Center (CERT/CC) is located at the Software Engineering Institute (SEI), a federally funded research and development center at Carnegie Mellon University in Pittsburgh, Pennsylvania. It is charged with coordinating communication among experts during computer security emergencies and helping to prevent future incidents. CERT employees study Internet security vulnerabilities, handle computer security incidents, publish security alerts, research long-term changes in networked systems, develop information and training to help organizations improve security at their sites, and conduct an ongoing public awareness campaign. The number of security problems reported to CERT increased twenty-five-fold between 1997 and 2001, as shown in Table 3-1. There are many reasons for this growth, as outlined in the subsequent paragraphs.

Table 3-1   Number of incidents reported to CERT

Year	Number of Incidents Reported
2001	52,658
2000	21,756
1999	9,859
1998	3,734
1997	2,134

Source: CERT Web site at www.CERT.org/stats, accessed on May 29, 2002.

Increasing Complexity Increases Vulnerability

Our computing environment has become enormously complex, as networks, computers, operating systems, applications, switches, routers, and gateways that are driven by hundreds of millions of lines of code are interconnected. This environment continues to increase in complexity day by day. The number of possible entry points to a network expands continually as more and more devices are added to it, thus increasing the possibility for security breaches to take place.

Higher Computer User Expectations

Today, time means money and the faster a computer user can solve a problem, the sooner he or she can be productive. As a result, computer help desks are under intense pressure to provide fast response to computer users' questions such as "How can I access this Web page?" or "Will you please assign me a new password; I forgot my old one." Under duress, help desk personnel sometimes forget to verify the user's identity or check if they are authorized to perform the requested action. In addition, even though they have been warned against doing so, some computer users will share their login ID and password with other coworkers who have forgotten their own passwords. In this manner, it is possible for workers to gain access to information systems and data for which they are not authorized.

Expanding and Changing Systems Introduce New Risks

We have moved from an era of standalone computers where critical data was stored on an isolated mainframe computer in a locked room to a network era where personal computers connect to networks with millions of other computers, all capable of sharing information with one another. Businesses have moved quickly into e-commerce, mobile computing, collaborative work groups, and global business. They rely on information technology to help them get there. However, it is increasingly difficult for any organization to keep up with the pace of technological change, successfully perform an ongoing assessment of new security risks, and implement approaches for dealing with them.

Increased Reliance on Commercial Software with Known Vulnerabilities

U.S companies increasingly rely on commercial software with known vulnerabilities. Even when those vulnerabilities are exposed, many corporate IT organizations prefer to use already-installed software "as is" rather than implement security fixes that will make the software harder to use or eliminate "nice-to-have" features. For example, in May 2000, Microsoft announced that it would make available a patch to Outlook 98 and 2000 (a contact, calendar, and e-mail program) so that the program would no longer accept certain types of programs as an attachment. This would stop the easy distribution of so called worm programs often contained in a type of program called VBScript. In addition, Outlook would no longer allow other programs to access its Address book unless the user explicitly agreed to each request. This would have stopped worms from automatically sending themselves to everyone in an infected computer's Address book, but it would have added a manual step to legitimate activities, such as synchronizing Outlook with a Palm handheld device. In response to strong negative feedback from corporate information technology managers, Microsoft decided to scale back on the tougher security planned for a new version of Outlook.[2]

In such an environment of increasing complexity, higher user expectations, expanding and changing systems, and increased reliance on software with known vulnerabilities, it is no wonder that the number of security incidents is increasing dramatically.

Types of Attacks

Security incidents can take many forms, but one of the most frequent kinds of incident is an attack on a networked computer from an outside source. There are numerous types of attacks with, unfortunately, new types of attacks being invented all the time. Currently, most attacks fall into one of the following categories: virus, worm, Trojan horse, and denial-of-service attack.

Virus

"Computer virus" has become an umbrella term for many types of malicious code. Technically, a **virus** is a piece of programming code usually disguised as something else that causes some unexpected and usually undesirable event. Often a virus is attached to a file so that when the infected file is opened, the virus executes. Other viruses sit in a computer's memory and infect files as the computer opens, modifies, or creates the files. Most viruses deliver a "payload" or malicious act. For example, the virus may be programmed to display a certain message on the computer's display screen, delete or modify a certain document, or reformat the hard drive.

A true virus does not spread itself from computer to computer. In order to propagate to other machines, it must be passed on to other users through infected e-mail document attachments, programs on diskettes, or shared files. In other words, it takes action on the part of the computer user to spread a virus.

Macro viruses have become the most common and easily created viruses. These viruses use an application macro language (such as Visual Basic or VBScript) to create programs that infect documents and templates. After an infected document is opened, the virus is executed and it infects the user's application templates. Macros can insert unwanted words, numbers, or phrases into documents or alter command functions. After a macro virus infects a user's application, it can embed itself in all future documents created with the application.

Worm

A computer virus requires the user to actively pass an infected file to another user in order to be spread. **Worms** are also harmful computer programs, but they differ from viruses because they have the ability to propagate without human intervention. For example, in 2000, the infamous ILOVEYOU worm automatically e-mailed itself to all contacts in the user's Microsoft Outlook Address book after the user opened an infected e-mail attachment. It took two days to develop and distribute a patch to overcome the ILOVEYOU bug. During this brief time, the worm had copied and distributed itself to some ten million computers. The cost to repair the damage done was estimated to be in the billions.[3] Its victims included the British Parliament, the United States Senate, and the Pentagon. The source of the bug was an embittered student in the Philippines; the program began its life as a project that was rejected by a teacher.

The ILOVEYOU bug was not the only costly worm to appear in recent years. The cost to repair the damage done by each of the Code Red, SirCam, and Nimda worms (all released in 2001) exceeded $500 million.

Trojan Horse

A **Trojan horse** is a program that gets secretly installed on a computer, planting a harmful payload that can allow the hacker who planted it to do such things as steal passwords or spy on users by recording keystrokes and transmitting them to a third party. Users are often tricked into installing Trojan horses. For example, a Trojan horse might arrive in an e-mail as a program described as a computer game. When users receive the mail, they may be enticed by the description of the game to install it. Although it may in fact be a game, it may also be taking other action that is not readily apparent to the users, such as deleting files or mailing sensitive information to the attacker. A **logic bomb** is a type of Trojan horse that executes when specific conditions occur. Logic bombs can be triggered by a change in a particular file, the typing of a specific series of keystrokes, or by a specific time or date.

An example of a Trojan horse is the Cute Trojan horse, which allows attackers to take over their victims' machines and destroy firewall and security programs. The Trojan horse arrives attached to an e-mail message with the subject line "Thoughts..." The message says, "I just found this program, and, i dont know why... but it reminded me of you. check it out." An attachment called Cute.exe accompanies the e-mail message. If the user opens the attachment, a Trojan horse is unleashed that will allow attackers to take control of the computer so that they can send e-mail and instant messages and access, move, copy, or delete files from the infected machine. The program also looks for security programs (such as antivirus and firewall programs) in memory and destroys them if found.[4]

Denial-of-Service Attacks

A **denial-of-service attack** is one in which a malicious hacker takes over computers on the Internet and causes them to flood a target site with demands for data and other small tasks. A denial-of-service attack does not involve a break-in at the target computer; instead, it just keeps the target machine so busy responding to the stream of automated requests that legitimate users cannot get in—the Internet equivalent of dialing a telephone number repeatedly so that everyone else trying to get through hears a busy signal.

The software to initiate a denial-of-service attack is simple to use and readily available at hacker sites. A tiny program is downloaded from the attacker's computer to dozens, hundreds, or even thousands of computers all over the world. Based on a command by the attacker or a preset time, these computers (called **zombies**) go into action, each one sending a simple request for access to the target site, again and again and again—dozens of times a second.

The zombies are often programmed to put false return addresses on the packets they send out (known as **spoofing**) so that the sources of the attack are obscured and cannot be identified and turned off. This fact actually provides an opportunity to prevent such attacks. Internet Service Providers (ISPs) can prevent incoming packets with false IP addresses from being passed on by a process called **ingress filtering**. Corporations with an Internet connection can ensure that spoofed packets don't leave their corporate network using a process called **egress filtering**. However, such checking of addresses takes a tremendous amount of Internet router processing power. As the number of packets increases, more and more processing capacity is required to check the IP address on each packet. Companies would have to deploy faster and more powerful routers and switches to maintain the same level of performance. As a result, few ISPs or corporations perform this checking. Such capability may be built into the next generation of network equipment. Considerable cost is involved in making such upgrades to routers and other network equipment.

The zombies involved in a denial-of-service attack have been very seriously compromised and are left with more enduring problems than their target. As a result, zombie machines need to be carefully inspected to ensure that the attacker software is completely removed from the system. In addition, system software must be reinstalled from a reliable backup to reestablish the system's integrity, and an upgrade or patch must be implemented to eliminate the vulnerability that originally allowed the attacker to enter the system.

In February 2000, over a period of two days, a series of denial-of-service attacks hit several of the then most frequently visited sites on the Web—Yahoo, Buy.com, Amazon.com, CNN.com, E*trade, and Excite. It now appears that a single teenager was easily able to cripple these large and popular sites. The attacks on these Web sites raised awareness of the vulnerability of e-commerce sites to hacking and the resultant business disruption. This realization contributed to the drop in enthusiasm that e-commerce and dot-com Web sites experienced in 2000 and 2001. In May 2001, the CERT Web site itself was bombarded by a denial-of-service attack.

Perpetrator Types

The perpetrators of computer crime are the same as they are for any other type of crime—thrill seekers wanting to take on a challenge, common criminals looking for financial gain, industrial spies trying to gain a competitive advantage, and terrorists seeking to cause destruction in order to bring attention to their cause. Each type of perpetrator has different objectives and access to varying resources, and each is willing to accept different levels of risk to accomplish the objective. Knowing these parameters for each set of likely attackers is the first step toward establishing effective countermeasures. Table 3-2 summarizes these parameters and the subsequent paragraphs discuss each type of perpetrator in more detail.

Table 3-2	Classification of perpetrators of computer crime			
Type of Perpetrator	Objective	Resources Available to Perpetrator	Level of Risk Taking Acceptable to Perpetrator	Frequency of Attack
Hacker	Test limits of system and gain publicity	Limited	Minimal	High
Cracker	Cause problems, steal data, and corrupt systems	Limited	Moderate	Medium
Insider	Financial gain and disrupt company's information systems	Knowledge of systems and passwords	Moderate	Low
Industrial spy	Capture trade secrets and gain competitive advantage	Well-funded and well-trained	Minimal	Low
Cybercriminal	Financial gain	Well-funded and well-trained	Moderate	Low
Cyberterrorist	Cause destruction to key infrastructure components	Not necessarily well-funded or well-trained	Very high	Low

Hacker and Cracker

Hackers are individuals who test the limitations of systems out of intellectual curiosity—to see whether they can gain access and how far they can go. They have at least a basic understanding of information systems and security features, and much of their motivation comes from a desire to learn even more. A common profile for today's hacker is one of a male who is in his mid-twenties or younger, has lots of spare time, has minimal financial resources, and is a social outsider. Some hackers are smart and talented. Many are technically inept and are referred to as **lamers** or **script kiddies** by those more skilled. Surprisingly, there is a wealth of resources available for hackers to use to hone their skills—online chat groups, Web sites, downloadable hacker tools, and even hacker conventions (such as Defcon, an annual gathering in Las Vegas).

Many hackers operate under the mistaken belief that it is ethical to break into a system and just "look around." However, such action can cause inadvertent damage and the system owner, upon finding out the system has been hacked, can no longer trust its integrity. For example, in the summer of 2000, the University of Washington Medical Center was hacked and the perpetrator downloaded files containing information on some 5,000 patients. The perpetrator said that he considers himself an ethical hacker who simply wanted to expose the vulnerability of the hospital's network.[5] However, such a break-in can cause patients to lose faith in a hospital's ability to safeguard sensitive data, and it can cause irreparable damage to its reputation, regardless of how successfully the organization subsequently implements new security measures.

Cracking is hacking that is clearly a form of criminal activity. **Crackers** break into other people's networks and systems, deface Web pages, crash computers, spread harmful programs or hateful messages, and write scripts and automatic programs that let other people do these things. In January 2000, the server of CD Universe, a Web site for CDs, movies, and games, was attacked by a cracker named Maxus who tried to extort $100,000 after copying 300,000 customer credit card numbers. When the company refused to deliver the money, the cracker posted 25,000 of the card numbers on the Internet. In December 2000, Creditcards.com, a company that provides credit card information for consumers and small-to-mid-sized merchants, was hit by a similar extortion attempt. Credit card theft is so common that public online exchanges have formed where stolen numbers are bought, sold, and traded.

The use of a stolen credit card is a type of **criminal fraud**, which is defined as the obtaining of title to property through deception or trickery. Fraud is a form of corruption that all companies need to be aware of and work directly to prevent. Read the Legal Overview to learn more about fraud.

LEGAL OVERVIEW

FRAUD

Obtaining title to property through deception or trickery constitutes the crime of criminal fraud. A fraudster may seek tangible gain (money or property) or intangible gains such as promotion, power, or influence.

To prove fraud, the following elements must be shown:

1. The wrongdoer made a false representation of material fact.
2. The wrongdoer intended to deceive the innocent party.
3. The innocent party justifiably relied on the misrepresentation.
4. The innocent party was injured.

Companies are exposed to a wide range of fraud risks, including diversion of company funds, theft of assets, fraud connected with bidding processes, invoice and payment fraud, computer fraud, and credit card fraud.

Not surprisingly, fraud that occurs within an organization is usually due to weaknesses in an organization's system of internal control procedures. As a result, most frauds are discovered by chance and by outsiders (from a tip from an outsider, through resolving payment issues with contractors or suppliers, or during a change of management) rather than through control procedures. Often frauds involve some form of **collusion** or cooperation between an employee and an outsider to commit fraud. For example, an employee in accounts payable may engage in collusion with a person in a supplier's accounts receivable organization. Each time the supplier submits an invoice, the employee in accounts payable adds $1000 to the amount approved for payment. The inflated payment is received by the supplier and the two split the extra money.

Identity fraud is the fastest growing form of fraud in the United States. The use of computers and the Internet have made identity fraud so much easier that credit reporting firms say the number of identity fraud cases has increased from 10,000 in 1990 to over 500,000 cases per year today. Taking on the identity of another can be easy and extremely lucrative, enabling the fraudster to use the victim's credit cards, siphon off money from bank accounts, and even obtain Social Security benefits. The victim is left with a credit history in shambles and may face thousands of dollars in costs trying to straighten out the mess.

Congress passed the Identity Theft and Assumption Deterrence Act of 1998 to fight identity fraud. This act criminalizes identity fraud, making it a federal felony punishable by a prison sentence ranging from three to twenty-five years. The act also appoints the Federal Trade Commission (FTC) to help victims restore their credit and erase the impact of the imposter. There have been numerous convictions of people under this act. For example, an individual was indicted on bank fraud charges for obtaining names, addresses, and Social Security numbers from a Web site and using the data to apply for a series of car loans over the Internet.[6]

Malicious Insiders

The number one security concern for companies is the malicious insider—an ever-present adversary. Indeed, it is estimated that more than 70 percent of network intruders come from inside the organization.[7] Insiders are not necessarily employees; they can be consultants and contractors as well. Nor do they need to be employees in IT-related positions; they may just be experienced IT users. Their risk tolerance ranges from low to high, depending on whether they are motivated by financial gain, revenge on their employers, or publicity.

Malicious insiders are extremely difficult to detect or stop because they're often authorized to access the very systems they abuse. Although insiders are less likely to attack systems than are outside hackers/crackers, the company's systems are far more vulnerable to them. Most computer security measures are designed to stop the external attacker, but they are nearly powerless against insiders. Insiders have knowledge of individual systems, often including the procedures to gain access to login IDs and passwords. Insiders know how the systems work and where the weak points are. Their knowledge of the organizational structure and security procedures helps them avoid any investigation of their actions.

Industrial Spies

Industrial spies use illegal means to obtain trade secrets from the competitors of the firm for which they are hired. Trade secrets are actually protected by the Economic Espionage Act of 1996, which makes it a federal crime for someone to use a trade secret for his own benefit or for the benefit of another. The theft of trade secrets is most often perpetrated by insiders, such as disgruntled employees and ex-employees.

Competitive intelligence uses legal information-gathering techniques to obtain information that is available to the public. Participants gather and analyze information from financial reports, trade journals, public filings, and printed interviews with company officials. Industrial espionage involves using illegal means to obtain information that is not available to the public. Participants might place a wiretap on the phones of key company officials, bug a conference room, or break into a research and development facility to steal confidential test results. An unethical firm may spend a few thousand dollars to hire an industrial spy to steal trade secrets that can be worth one hundred times that amount. The industrial spy avoids taking risks that would expose his employer—the employer's reputation (an intangible but valuable item) will be damaged considerably if it is caught spying on the competition.

Industrial espionage can involve the theft of new product design, production, or marketing information. For example, in 1998, an employee of Wright Industries, a Gillette supplier in Nashville, was convicted of industrial espionage for stealing highly confidential product information belonging to the razor manufacturer and distributing it to its competitors, costing Gillette $1.5 million in damages.[8] Sometimes industrial espionage involves the theft of simple materials that are nonetheless essential to a firm's success. Fast-food chain In-N-Out Burger collected $250,000 in 2002 from an individual whom they claimed had conducted industrial espionage to copy the chain's operating methods by stealing uniforms, training material, and financial reports.[9]

Cybercriminals

Information technology provides a new and highly profitable venue for **cybercriminals**. They hack into corporate computers and steal money, often by transferring money from one account to another to another—leaving a hopelessly complicated trail for law enforcement officers to follow. Cybercriminals also engage in all forms of computer fraud—stealing credit card numbers, personal identities, and cell phone IDs and reselling them. They can spend large sums of money to buy the technical expertise and access needed to perpetrate this fraud using unethical insiders willing to sell their knowledge of passwords and access procedures.

The use of stolen credit card information is a favorite ploy of computer criminals. Online credit card fraud rates vary greatly from roughly .06 percent to 4 percent of online credit card transactions, with certain kinds of merchants having much higher fraud rates than others. For example, rates are highest for merchants that sell downloadable software or expensive items such as electronics and jewelry (because of their high resale value). In March 2001, the FBI said that over one million credit card numbers were stolen from more than 40 businesses in 20 states by cybercriminals in Russia and the Ukraine. The criminals used automated software tools to scan the Internet for computers that had not applied fixes to correct well-known vulnerabilities in certain Microsoft software.[10]

Credit card companies are so concerned about making consumers feel safe shopping online that many are marketing new and exclusive zero-liability programs (see Case #2). This is in spite of the fact that the Fair Credit Billing Act limits consumer liability to only $50 worth of unauthorized charges. The cost of these fraudulent charges is borne by the bank issuing the credit card or by the merchant. For credit card purchases in retail stores, the customer presents a credit card, the clerk swipes it through an electronic reader, and the customer signs the credit slip. When a charge is made fraudulently in a retail store, the bank issuing the credit card must pay the fraudulent charges. For credit card transactions over the Internet, the Web merchant absorbs the cost.

A high rate of disputed transactions, known as **chargebacks**, can greatly reduce a Web merchant's profit margin. However, the permanent loss of revenue caused by losing its customers' trust has far more impact than the cost associated with fraudulent purchases and bolstering security. Indeed, most companies are afraid to admit publicly that they've been hit by online fraud or hacker intrusions because they don't want to hurt their reputations.

An example of acting ethically in communicating with the public about security problems involves the software vendor PDG Software, which is a provider of software to support consumer to business e-commerce. In April 2001, PDG revealed that computer criminals had figured out a way to break into its software and raid customer accounts for credit card information. At risk were all customer records in nearly 4,000 Web sites operated by PDG customers. Within hours of becoming aware of the situation, PDG provided a free patch for all merchants using the software. They also notified all customers of the patch's availability and alerted them to the seriousness of the situation and the effects that not installing the patch could have upon their system. Fortunately, most merchants using the PDG software immediately upgraded their systems and were not seriously affected. The following day, PDG was notified that some merchants had experienced unauthorized access to their sites and PDG decided to go to the FBI, which issued a public warning directed at the software manufacturer's customers.

Some PDG customers had failed to implement the fix. As a result, more than one hundred consumers had thousands of dollars fraudulently charged on their credit cards. However, the number of consumers affected was greatly reduced by the fast action of PDG. In the subsequent months, PDG received many new customers who indicated that they were impressed with the manner in which PDG had handled this problem.[11]

There are many approaches to reducing the potential for online credit card fraud. Most e-commerce Web sites use some form of encryption technology to protect information as it moves from the consumer to the Web site. Some also employ address verification by matching the address submitted online to the one the issuing bank has on file. However, this presents the risk that the merchant may inadvertently throw out legitimate orders with different shipping addresses than those listed in the files, such as when a consumer places a legitimate order but requests shipment to a different address because it is a gift. Another security technique is to ask for a card verification value (CVV), which is the three-digit number above the signature panel on the back of a credit card. This makes it impossible for fraudsters who have the credit card number but not the card itself to make purchases. An additional security option is transaction-risk scoring software, which keeps track of a customer's historical shopping patterns and notes deviations from the norm. Assume that you have never been to a casino and your credit card comes up at Caesar's Palace at 2 A.M. Your transaction-risk score would go up dramatically, so much so that the transaction might be declined.

Some card issuers are implementing **smart cards** that contain a memory chip that is updated with encrypted data every time the card is read. To use a smart card for online transactions, a consumer must purchase a card reader that attaches to his or her personal computer and enter a personal identification number to gain access to the account. Although smart cards are used widely in Europe, they are not widely used in the United States because of the changeover costs for merchants.

Cyberterrorists

A **cyberterrorist** is someone who intimidates or coerces a government or organization to advance his or her political or social objectives by launching computer-based attacks against computers, networks, and the information stored on them. Such attacks could be accomplished by the sending of a virus or worm or through the launching a denial-of-service attack. Because of the Internet, attacks can easily originate from foreign countries, making detection and retaliation much more difficult.

The United States government considers the potential threat of cyberterrorism serious enough that it established the National Infrastructure Protection Center in February 1998. The mission of this branch of the FBI is to serve as the United States government's focal point for threat assessment, warning, investigation, and response for threats or attacks against our country's critical infrastructures. These infrastructures include telecommunications, energy, banking and finance, water, government operations, and emergency services. For instance, targets might include telephone-switching systems, the electric power grid that serves major portions of a geographic region, or the air traffic control center that ensures airplanes can take off or land safely. Successful cyberattacks on such targets could cause widespread and massive disruptions to the normal functioning of society. Some computer security experts believe that cyberterrorism attacks could be used to further complicate matters following a major nuclear, biological, or chemical act of terrorism.[12]

Cyberterrorists seek to cause harm rather than to gather information, and they use techniques that result in outright destruction or disruption of services. They are extremely dangerous, consider themselves to be in a personal state of war, have a very high acceptance of risk, and seek negative publicity. Despite their extremism, terrorist groups pose only a limited threat to information systems because they believe that "bombs still work better than bytes," and most operate on a very limited budget.[13] There are exceptions, with members of some terrorist organizations being well-organized, well-trained, and well-supported, but the majority of groups don't have good organization or access. However, the threat from cyberterrorist attacks can be expected to increase as younger, computer-savvy terrorists rise in the ranks of terrorist organizations.[14]

On March 30, 2000, FBI Houston conducted a search warrant on a residence of an individual who allegedly created a computer worm that seeks out computers on the Internet. The worm causes the hard drives of randomly selected computers to be erased. The computers whose hard drives are not erased actively scan the Internet for other computers to infect and force the infected computers to use their modems to dial 911. Because each infected computer can scan approximately 2,550 computers at a time, this worm had the potential to create a denial-of-service attack against the emergency 911 system. Although the motives of the individual who created the worm were not clear, the potential for serious impact on the emergency response systems was.[15]

Other examples of cyberterrorism abound. During the Kosovo conflict in 1999, NATO computers were blasted with e-mail with harmful attachments and hit with denial-of-service attacks by cyberterrorists protesting the NATO bombings in Kosovo. Lucent Technologies fought off a denial-of-service attack on its Web site in 2000 that apparently was launched by a group protesting the amount of business the company does with Israel.[16] Following the April 2001 collision between a United States spy plane and a Chinese jet fighter, cyberterrorism attacks by activists in both countries brought down or defaced hundreds of Web sites.

REDUCING VULNERABILITIES

The security of any system or network is a combination of technology, policy, and people. Thus a proactive IT security group must perform a wide range of activities to ensure implementation of effective security measures. A strong security program begins with an assessment of the threats to the organization's computers and network and the identification of those actions that must be taken to address the most serious vulnerabilities.

Education for users about the risks involved and the actions they must take to prevent a security incident is a key part of any successful security program. The IT group must lead the effort to put into place security policies and procedures along with hardware and software tools to help prevent security breaches. However, no security system is perfect, so monitoring systems and procedures must also be implemented to detect a possible intrusion. If an intrusion occurs, there must be a clear reaction plan that addresses notification, evidence protection, activity log maintenance, containment, eradication, and recovery. These activities are spelled out in more detail below.

Risk Assessment

A **risk assessment** is an organization's review of potential threats to its computers and network and the probability of those threats occurring. Its goal is to identify those investments in time and resources that will best protect the organization from its most likely and serious threats. No amount of resources can guarantee a perfect security system, so organizations frequently have to balance the risk of a security breach with the cost of preventing one. The concept of **reasonable assurance** recognizes that managers must use their judgment to ensure that the cost of control does not exceed the benefit to be obtained by its implementation or the possible risk involved. Table 3-3 illustrates a risk assessment for a hypothetical organization.

A completed risk assessment identifies the most dangerous threats to a company and thus helps focus security efforts where there is the highest payoff. For each risk area, the estimated probability of such an event occurring is multiplied by the estimated cost of a successful attack. The product is the expected cost impact for that risk area. An assessment is then made of the current level of protection against that event occurring. The level of protection can be poor, good, or excellent. The risk areas with the highest estimated cost and with the poorest current level of protection are the areas where security measures need to be improved.

Table 3-3	Risk assessment for hypothetical company				
Risk	Estimated Probability of Such an Event Occurring	Estimated Cost of a Successful Attack	Probability × Cost = Expected Cost Impact	Assessment of Current Level of Protection	Relative Priority to be Fixed
Denial-of-service attack	80%	$500,000	$400,000	Poor	1
E-mail attachment with harmful worm	70%	$200,000	$140,000	Poor	2
Harmful virus	90%	$50,000	$45,000	Good	3
Invoice and payments fraud	10%	$200,000	$20,000	Excellent	4

The United States government conducts frequent security assessments of its agencies—usually with disappointing results. In 2000, a congressional subcommittee assessing the vulnerability of government computer systems to attacks from terrorists and hackers gave half of the 54 federal departments and agencies less than satisfactory grades.[17] Also in 2000, the security practices of the Federal Aviation Administration (FAA) were audited by the General Accounting Office and the FAA was strongly criticized for failing to do background security checks on many of its contract workers—including some who were hired to conduct penetration testing of the agency's computer systems.[18]

Clearly, it takes action in addition to assessment in order to plug holes in the security defenses of information systems. The following sections outline the necessary steps.

Establish A Security Policy

A **security policy** defines the security requirements of an organization and describes the controls and sanctions to be used to meet those requirements. An important piece of any policy is the delineation of the responsibilities and expected behavior of members of the organization. A security policy outlines *what* needs to be done, but not *how* to do it. It should refer to procedure guides instead of outlining the procedures.

Wherever possible, automated system policies should be implemented that mirror an organization's written policies. These policies can often be put into practice using the configuration options in a software program. For example, if a written policy states that passwords must be changed every 30 days, then all systems should be configured to enforce this policy automatically. There are always going to be trade-offs when applying system security restrictions. Often the trade-off is between ease of use and increased security; however, when the decision is made in favor of ease of use, there is sometimes a resulting increase in security incidents.

The use of e-mail attachments is a critical issue to be addressed in a security policy. Sophisticated attackers can try to penetrate a network via e-mail and e-mail attachments, regardless of the existence of a firewall and other security measures. As a result, many companies have implemented a policy of blocking all incoming mail that has executable file attachments. This greatly reduces their vulnerability. Some companies have an e-mail policy that allows employees to receive and open e-mail with attachments, but only if the e-mail is expected and from someone known by the recipient. However, such a policy can be risky because many worms use the Address book of their victims to generate e-mails to a target audience known by the victims.

Educate Employees, Contractors, and Part-Time Workers

Employees, contractors, and part-time workers must be educated about the importance of security so that they will be motivated to understand and follow the security policy. Often this can be accomplished by a discussion of recent security incidents that had an impact on the organization. Users must understand that they are a key part of the security system and that they have certain responsibilities. For example, users must help protect an organization's information systems and data by doing the following:

- Guarding their passwords to protect against unauthorized access to their accounts
- Not allowing others to use their passwords
- Applying strict access controls (file and directory permissions) to protect data from disclosure or destruction
- Reporting all unusual activity to the organization's IT security group

Prevention

No organization can ever be 100 percent secure from attack. The key is to implement a layered security solution to make breaking into an organization harder than the attacker is willing to work, so that if an attacker breaks through one layer of security, there is still another layer to overcome. Here are a number of items that can be implemented to provide the desired layers of security.

Install a Corporate Firewall

Installation of a corporate firewall is the most common security precaution taken by businesses. As discussed in Chapter 2, a **firewall** stands guard between your organization's internal network and the Internet and limits access into and out of your network based on the organization's access policy. Whatever Internet traffic is not explicitly permitted to pass through to the internal network is denied entry.

Similarly, internal network users can be blocked from gaining access to certain Web sites based on content such as sex, violence, and so on. The firewall can also block instant messaging, access to newsgroups, and other Internet activities.

Installing a firewall can lead to another serious security issue—complacency. A firewall does nothing to protect a Web site from a denial-of-service attack. A firewall also cannot prevent a worm from entering the network as an e-mail attachment. Most firewalls are configured to allow e-mail and benign-looking attachments to reach their intended recipient.

Although firewalls are frequently used to improve security in handling e-commerce transactions, they can also slow down communications between shoppers and online order-processing systems. For example, during the 2001 Christmas holiday season, some online retailers kept holiday shoppers waiting nearly 15 seconds while their Web transactions were processed. At least one company was able to circumvent this problem, though: JCrew.com, the online unit of J. Crew Inc., was able to process transactions in an average of 4.5 seconds due to a new firewall system that the retailer had its Web hosting provider deploy prior to the start of the holiday rush. The new firewall system cost $7,450 per month—40 percent more than the cost of JCrew.com's previous firewall offering. However, it was well worth the increased ability to process orders.[19]

Table 3-4 lists some of the firewall software used to protect home personal computers. Typically, the software sells for $30 to $60 for a single user license.

Table 3-4 Popular personal computer firewall software products	
Software	**Vendor**
Norton Personal Firewall	Symantec
Tiny Personal Firewall	Tiny Software
BlackIce Defender	Network Ice Corporation
ZoneAlarm Pro	Zone Labs
Personal Firewall	McAfee
Personal Firewall Pro	Sygate

Install Anti-Virus Software on Personal Computers

Anti-virus software should be installed on each user's personal computer to regularly scan a computer's memory and disk drives for viruses. To find such a virus, anti-virus software scans for a specific sequence of bytes, known as the **virus signature**. If it finds a virus, the anti-virus software informs the user and may clean, delete, or quarantine any files, directories, or disks affected by the malicious code. Good anti-virus software checks vital system files when the system is booted up; monitors the system continuously for virus-like activity; scans diskettes; scans memory when a program is run; checks programs when they are downloaded; and scans e-mail attachments before they are opened. Two of the most widely used anti-virus software products are Norton Antivirus from Norton and Dr. Soloman's Antivirus from McAfee.

CERT has long served as a clearinghouse for news on new viruses, worms, and other computer security topics. CERT estimates that over 99 percent of the major attacks the team analyzes use already known virus or worm programs.[20] Thus, it is crucial that anti-virus software be continually updated with the latest virus detection information, called **definitions**. In most corporations, it is the responsibility of the network administrator to monitor the network security providers' Web sites at least once a week and download updated anti-virus software as needed.

Implement Safeguards Against Attacks by Malicious Insiders

In a June 2001 online survey, 57 percent of 548 corporate security managers said that their worst security breaches were from corporate users accessing information for which they were not authorized. The next biggest problems were those created by user accounts left active after employees left the company.[21] To reduce the threat of attack by malicious insiders, the prompt deletion of the computer accounts, login IDs, and passwords of departing employees must be a high priority.

Organizations also need to carefully define the roles of their employees with a goal of proper separation of responsibility so that a single individual is not responsible for accomplishing a task that has high security implications. For example, it would not make sense to establish a role where an employee had the power to both initiate a purchase order and approve an invoice for the payment of that purchase order. Such an employee would be in a position to place a large dollar volume of orders with a "friendly vendor," approve the invoices for payment, and then disappear from the company to split the money with the vendor. In addition to separation of duties, many organizations frequently rotate people in sensitive positions to prevent potential insider crimes.

Another important safeguard is to create roles and user accounts so that users have the authority to perform their responsibilities and no more. For example, members of the finance department will have separate authorization from members of human resources. An accountant should not be able to review the pay and attendance records of an employee, and a member of human resources should not know how much was spent to modernize a piece of equipment. Even within a given department, not all members should be given the same capabilities. Within the finance department, some users may have the ability to approve invoices for payment, but not all users will. An effective administrator will identify the similarities among users and create roles associated with the groupings.

Address the Ten Most Critical Internet Security Threats

The majority of successful attacks on computer systems via the Internet can be traced to exploitation of one of a small number of technical security flaws. Top security experts from industry, government, and academia have developed a list of the top ten Internet security flaws that system administrators must eliminate to avoid becoming an easy target. The list, along with the actions needed to rid systems of these vulnerabilities, can be found at *www.sans.org/topten.htm*. The top-ten list is updated periodically.

The actions required to address the top-ten items are usually quite technical in nature. For example, the number one item on the top-ten list for June 25, 2001, involved making technical adjustments and an upgrade to the Berkeley Internet Name Domain (BIND) package. This package is the most widely used implementation of Domain Name Service (DNS), which is the means by which you locate systems on the Internet by name (such as *www.sans.org*) without having to know specific IP addresses. Failure to take the recommended action makes it possible for intruders to erase the system logs and installed tools so that they are able to gain administrative access without being detected.

Verify Backup Processes for Critical Software and Databases

It is imperative to back up critical applications and data regularly. However, there have been many cases of organizations implementing inadequate backup processes and finding out that the backup was not sufficient to enable a full restoration of the original data. All backups should be created in such a way and with enough frequency to enable a full and quick restoration of data in the event that an attack destroys the original. This process should be tested to confirm that it really works.

Conduct Periodic IT Security Audits

A security audit evaluates whether an organization has a well-thought-out security policy in place and if it is being followed. For example, if a policy says that all users must change their passwords every 30 days, the audit will check how that policy is implemented and whether it is truly happening. The audit will also review who has access to what systems and data and what level of authority each user has. It is not unusual for an audit to discover that too many people have access to critical data and that many people have capabilities above those needed to perform their jobs. One of the results of a good audit is a list of items to be corrected to ensure that the security policy is being met.

A thorough security audit should also test the system safeguards to ensure that they are operating as intended. Such tests might include trying the default system passwords that are active when software is first received from the vendor. The goal of such a test is to ensure that all such "known" passwords have been changed.

Detection

Even with prevention items implemented, no organization is 100 percent secure from a determined attack. Thus, it is important for an organization to implement detection systems to catch any intruders in the act. Organizations often employ intrusion detection systems or a "honeypot" (lure) to minimize the impact of intruders.

Intrusion Detection Systems

An **intrusion detection system** monitors system and network resources and activities and, using information gathered from these sources, notifies the proper authority when it identifies a possible intrusion.[22] There are two fundamentally different approaches to intrusion detection—knowledge-based approaches and behavior-based approaches. Knowledge-based intrusion detection systems contain information about specific attacks and system vulnerabilities and watch for attempts to exploit these vulnerabilities. Examples include actions such as repeated failed login attempts, someone attempting to download a program to a server, or other symptoms of possible mischief. When such an attempt is detected, an alarm is triggered. With behavior-based intrusion detection systems, a model of normal behavior of the system and its users is developed from reference information collected by various means. The intrusion detection system compares current activity to this model and generates an alarm if a deviation is observed. Examples include unusual traffic at odd hours, a user known to be in the human resources organization accessing an accounting program that he or she never used before, or other actions that are abnormal.

Honeypot

The idea of a network-based **honeypot** is to provide would-be hackers with fake information about a network by means of a decoy server to confuse them, trace them, or keep a record for prosecution. The honeypot is well-isolated from the rest of the network and is capable of extensively logging the activities of intruders. The concept is still new, but a few companies have developed honeypots.[23]

Typically, reconnaissance probes occur prior to a real attack because these probes enable an attacker to obtain necessary information about the network resources he or she wishes to attack. The ActiveResponse honeypot from ForeScout Technologies identifies all such reconnaissance activity, and when the network responds back to the potential attacker, it provides fictitious data that mimics exactly the type of information that the potential attacker would get back from legitimate network resources. In the future, if network traffic based on the tagged information is observed, the honeypot recognizes that the attacker has returned to mount an actual break-in attempt. Action is initiated to block the session and capture information about the attack, such as dates, time, and all keystrokes entered.

Response

An organization should be prepared for the worst, which is a successful attack that defeats its defenses and causes damage to data and information systems. A response plan should be developed well in advance of any incident and be approved by both the legal department and senior management. An already-developed response plan helps keep the incident under technical and emotional control.

In the event of a security incident, the primary goal must be to regain control and limit damage, not to attempt to monitor or catch an intruder. System administrators have been known to take the discovery of an intruder as a personal challenge and to lose valuable time that should have been used in restoring data and information systems to normal.

Incident Notification

A key element of any response plan is to define who to notify and who *not* to notify. Who within the company needs to be notified, and what information does each person need to have? Under what conditions should the company be proactive and contact major customers and suppliers? How does the company inform them of a disruption in business processes without unnecessarily alarming them? When should CERT or even the FBI be contacted?

Most security experts recommend against giving out specific information about a compromise in public forums such as news reports, conferences, professional meetings, and online discussion groups. All parties working on the problem need to be kept informed and up-to-date. At the same time, the company needs to avoid use of systems connected to the compromised system. The intruder may be monitoring these systems and e-mail to learn what is known about the security breach.

Protect Evidence and Activity Logs

An organization should document all details of the security incident as it works to resolve it. Doing so will capture vulnerable evidence for use during any future prosecution and also will provide data to help during the incident eradication and follow-up phases. It is especially important to capture all system events, the specific actions taken (what, when, and who) and all external conversations (what, when and who) in a log book. Because this may become potential court evidence, a set of document handling procedures should be established using the legal department as a resource.

Incident Containment

Often it is necessary to act quickly to contain an attack and to keep a bad situation from becoming even worse. The response plan should clearly define the process for deciding if an attack is dangerous enough that critical systems should be shut down or disconnected from the network. How such decisions are made, how fast they are made, and who makes them are all elements of an effective response plan. In the event of a true security incident, as many defined procedures and policies as possible should be used.

Incident Eradication

Before the IT security group begins the eradication effort, they must collect and log all possible criminal evidence from the system. Creation of a disk image of each compromised system on write-only media for later study, and as evidence, can be very useful. Before performing the eradication, they will want to verify that all necessary backups are current, complete, and free of any virus. After virus eradication, they must create a new backup. Throughout this process, a log should be kept of all actions taken. This will prove helpful during the follow-up phase and will ensure that the problem does not reoccur.

Incident Follow-up

Of course, an essential part of follow-up is to determine how the organization's security was compromised so that it can be prevented from happening again. Often the fix is something as simple as getting a software patch from a product vendor. However, it is important to look deeper than the immediate fix and discover why the incident was able to occur. If a simple software fix could have prevented the incident, then why wasn't the fix installed *before* the incident occurred?

A review should be conducted after the incident to determine exactly what happened and to evaluate how the organization responded to the incident. One approach to doing this is to write a formal incident report that includes a detailed chronology of events and the impact of the incident. This will identify any mistakes made so that they are not repeated in the future. The experience gained from this incident should be used to update and revise the security incident response plan.

Creating a detailed chronology of all events will also document the incident in great detail for later prosecution. To this end, it is critical to develop an estimate of the monetary damage caused by the incident. Costs include loss of revenue, loss in productivity, and salaries of people working to address the incident. The cost to replace data, software, and equipment also needs to be considered.

Another important issue is how much effort should be put into trying to capture the perpetrator. If a Web site was simply defaced, it is easy to fix or restore the site's HTML (Hypertext Markup Language, the code that describes to your browser how a Web page should look). However, what if the intruders inflicted more serious damage, such as erasing proprietary program source code or the contents of key corporate databases? What if they stole company trade secrets? An expert cracker can conceal his or her identity so that tracking him or her down can take tremendous corporate resources over a long period of time.

The potential for negative publicity must also be considered. Discussing security attacks through public trials and the associated publicity not only has enormous potential public relations costs, but also real monetary costs. For example, a brokerage firm might experience the loss of many customers who learn of an attack and then think their money or their records aren't secure. Even if a company decides the negative publicity risk is worth it and goes after the perpetrator, it's possible that documents containing proprietary information that must be provided to the court could cause even greater security threats in the future.

Table 3-5 provides a manager's checklist for assessing an organization's ability to prevent and deal with an Internet security incident. "Yes" is the preferred answer for each question.

Table 3-5   Manager's checklist for evaluating an organization's readiness for an Internet security incident

Questions	Yes	No
Has a risk assessment been performed to identify those investments in time and resources that will protect the organization from its most likely and most serious threats?	____	____
Have those people involved in implementing security measures and senior management been educated about the concept of reasonable assurance?	____	____
Has a security policy been formulated and broadly shared throughout the organization?	____	____
Have automated systems policies been implemented that mirror the written policies?	____	____
Does the security policy address e-mail attachments with executable file attachments?	____	____

Table 3-5 Manager's checklist for evaluating an organization's readiness for an Internet security incident (continued)

Questions	Yes	No
Is there an effective security education program for employees, contractors, and part-time employees?	——	——
Has a layered security solution been implemented to prevent break-ins?	——	——
Has a firewall been installed?	——	——
Is anti-virus software installed on all personal computers?	——	——
Is the anti-virus software frequently updated?	——	——
Have precautions been taken to limit the impact of malicious insiders?	——	——
Are the accounts, passwords, and login IDs of former employees promptly deleted?	——	——
Is there a well-defined separation of employee responsibilities?	——	——
Are individual roles defined so that users have authority to perform their responsibilities and no more?	——	——
Is it a requirement to review the top-ten list of the most critical Internet security threats and take action to implement safeguards against them?	——	——
Has it been verified that backup processes for critical software and databases work correctly?	——	——
Are periodic IT security audits conducted?	——	——
Have intrusion detection systems been implemented to catch intruders in the act—both in the network and on critical computers on the network?	——	——
Has the installation of a honeypot to confuse and detect intruders been considered?	——	——
Has a comprehensive incident response plan been developed?	——	——
Has the plan been reviewed and approved by legal and senior management?	——	——
Does the plan address all of the following areas:		
Incident notification?	——	——
Protection of evidence and activity logs?	——	——
Incident containment?	——	——
Incident eradication?	——	——
Incident follow-up?	——	——

Summary

1. What are some key trade-offs and ethical issues associated with the safeguarding of data and information systems?

   Business managers, IT professionals, and IT users all face a number of ethical decisions regarding IT security, including the following: whether to pursue prosecution of computer criminals at all costs or maintain a low profile to avoid negative publicity; how much effort and money to spend implementing safeguards against computer crime (determining how safe is safe enough); what actions to take if one's firm produces software that contains defects that allow hackers to attack customers' data and computers; what tactics management should request employees take in gathering competitive intelligence; and what should be done if recommended computer security safeguards make life more difficult for customers and employees, resulting in lost sales and increased costs.

2. Why has there been a dramatic increase in the number of computer-related security incidents in recent years?

   The increasing complexity of the computing environment, higher user expectations, expanding systems, and increased reliance on software with known vulnerabilities have led to a twenty-five-fold increase in the number of reported IT security incidents over the past five years.

3. What are the most common types of computer security attacks?

   Currently, most incidents fall into one of the following categories: virus, worm, Trojan horse, and denial-of-service attack.

4. What are some of the characteristics of common computer crime perpetrators, including their objectives, available resources, willingness to accept risk, and frequency of attack?

   The perpetrators include the following: hackers who want to test the limits of a system, crackers who would wish to cause system problems, insiders who are seeking financial gain or revenge, industrial spies trying to gain a competitive advantage, cybercriminals looking for financial gain, and cyberterrorists seeking to cause destruction in order to bring attention to their cause. Each type of perpetrator has access to varying resources and is willing to accept different levels of risk to accomplish the objective. Knowing the parameters for each set of likely attackers is the first step toward establishing effective countermeasures.

5. What are the key elements of a multi-level process for managing security vulnerabilities based on the concept of reasonable assurance?

   A strong security program to protect against perpetrators begins with an assessment of the threats to the organization's computers and network. This assessment identifies those actions that must be taken to address the most serious vulnerabilities. Education of users to the risks involved and the actions they must take to prevent a security incident is a key part of any successful security program.

   The IT group must lead the effort to put into place security policies and procedures along with hardware and software tools to help prevent security breaches. No security system is perfect, so monitoring systems and procedures must also be implemented to detect a possible intrusion.

6. What actions must be taken in response to a security incident?

   If an intrusion occurs, there must be a clear response plan that addresses notification, protection of evidence and security logs, containment, eradication, and follow-up. Every attempt should be made to capture knowledge gained from one security incident to prevent or lessen the negative effects of a future security incident.

75

Review Questions

1. Identify four reasons for the growth in the number of IT security incidents.
2. Identify the key characteristics of six types of perpetrators of computer crime. Which type is considered the number one security risk for most companies?
3. What is fraud? What elements must be shown to prove fraud?
4. What is identity fraud? What are its effects? How prevalent is it?
5. What is the difference between a cyberterrorist and an industrial spy?
6. How is a virus different from a worm? How is a worm different from a Trojan horse?
7. What is CERT and what role does it play?
8. Describe a denial-of-service attack. How serious are such attacks?
9. Outline the key elements of a cost-effective program to reduce Internet security incidents.
10. Why must anti-virus software be constantly updated?
11. Why is it important to verify the backup process for critical software and databases?
12. Why is intrusion detection software needed? What types of software are there?
13. What activities are covered by a thorough security incident recovery plan?

Discussion Questions

1. Many people believe that a hacker is not a dangerous person and is not causing harm. What do you think?
2. What could ISPs and corporations do to reduce the threat of denial-of-service attacks? Why haven't these changes been implemented? Should Internet users or some federal agency demand that they be implemented? Why or why not?
3. How can installation of a firewall give an organization a false sense of security?
4. Some IT security people believe that their organizations should never fail to employ whatever resources are necessary to capture and prosecute computer criminals. Do you agree? Why or why not?
5. It is said that the more complex a system becomes, the more vulnerable it is to an attack. Why do you think this might be true?
6. Under what conditions do you think the corporate victims of computer crime should pursue their attackers to the full extent of their abilities and resources? What is the possible downside to such action?

What Would You Do?

1. You are the CFO of a mid-sized manufacturing firm. You have heard nothing but positive comments about the new CIO you hired three months ago. As you observe her outline what needs to be done to improve the firm's computer security, you are impressed with her energy, enthusiasm, and presentation skills. However, when she states that the total cost of the computer security improvements will be $300,000, your jaw drops. You had budgeted only $50,000 for this effort based on your assessment that security was okay because the firm had had no major incident. Several other items in your budget will either have to be dropped or trimmed way back to accommodate this project. In addition, the

$300,000 is above your spending authorization and will require that you get approval of the CEO. This will force you to defend the expenditure and you are not sure how to do this after minimizing the need in previous meetings.

You wonder if more than $50,000 of security is really required. How can you sort out what really needs to be done?

2. You have just been hired as an IT security consultant to "fix the security problem" at Acme United Global Manufacturing. The company has been hacked mercilessly over the last six months, with three of the attacks making the headlines of the business section for the negative impact they have had on the firm and its customers. You have been given 90 days and a budget of $1 million. Where would you begin, and what steps would you take to fix the problem?

3. You have learned that a friend of yours is developing a worm to attack the administrative systems at your institution of learning. The worm is "harmless" and will simply cause a message—Let's party!—to be displayed on all workstations connected to the computers on Friday afternoon at 3 P.M. By 4 P.M. the virus will erase itself and destroy all evidence of its presence. What would you do?

4. You are the vice-president of application development for a small, but rapidly growing software company that produces patient billing applications for doctors' offices. In working on the next release of your firm's one and only software product, a small programming glitch has been uncovered in the current release that, if left uncorrected, poses a security risk to its users. The probability of the problem being uncovered is low, but if exposed, the potential impact on your firm's 100 or so customers could be substantial— hackers could get access to private patient data and change billing records. The problem will be corrected in the next release, but you are concerned about what should be done for the users of the current release.

The problem has come at the worst possible time. The firm is seeking approval for a $10 million loan to raise enough cash to continue operations until revenue from the sale of its just-released product offsets expenses. In addition, the effort to communicate with users, to develop and distribute the patch, and to deal with any fallout will place a major drain on your small development staff so that the next software release will be delayed at least one month. You have a meeting with the CEO this afternoon; what course of action will you recommend?

Cases

1. Cybercrime: Even Microsoft Is Vulnerable

On October 27, 2000, Microsoft acknowledged that its security had been breached and that outsiders using a Trojan horse virus had been able to view source code for computer programs under development. A Microsoft spokesperson, Rick Miller, called the break-in "a deplorable act of industrial espionage." The incident was discovered by the company on October 25, 2000, and was reported to the FBI the next day. The attack, which is being investigated by the FBI, was believed at first to have initiated from St. Petersburg, Russia. Security experts believe that the attack could potentially have serious repercussions for Microsoft.

Customers as well as competitors of Microsoft were amazed that Microsoft was vulnerable. Most assumed that the world-renowned software leader had bulletproof security and was untouchable. The incident sent a loud wake-up call to organizations everywhere to step up security—if it can happen to Microsoft, it can happen to anyone.

Initially, Microsoft reported that the hacking had gone on over a six-week period. Microsoft later amended its initial reports and said the incident lasted only a week. Microsoft's corporate security officer said that the confusion stemmed from initial uncertainty over whether routine virus incidents in September were related.

There was also uncertainty over just what the hackers had seen or done. Initially, Microsoft said that the hackers were able to view, but not modify, source code of several major products. Later Microsoft amended this statement and said that the incident was not as serious as it originally feared—the intruder did not gain access to the source code of any strategic major products. However, the *Washington Post*, citing a source close to the investigation, reported that the targeted material was related to Microsoft's .NET strategy, its far-reaching plan to build the Internet into all its software.

One thing that does seem to be certain is that passwords to access Microsoft systems were stolen. Security analysts now believe that the passwords were sent to an e-mail account in China (initially reported to be St. Petersburg, Russia).

The attackers apparently used a worm known as QAZ to break into Microsoft's network. Once inside, the worm broadcasted its location to the cracker, who then took administrative control of the system without the user's knowledge. This enabled the attacker to do the same things the authorized user of the computer was permitted to do. It was also programmed to deliver passwords from one computer to another.

Experts said that an attack with a worm such as QAZ shouldn't have been possible if Microsoft had properly configured its firewall and anti-virus software and kept them updated. Anti-virus software vendors were familiar with QAZ and had updated their packages to detect QAZ months before the attack. A description of the worm was even posted on their Web sites, including steps users can take to protect themselves from QAZ.

One scenario of how this may have happened is that a Microsoft employee logged onto the company network from a home computer and inadvertently revealed system and network passwords to a hacker watching online. (The odds of a worm being downloaded on a home computer are much greater than on an office-based one, because home security is frequently less stringent and harder to monitor.) The intruders were then able to send e-mails to Microsoft computers laced with the QAZ worm program. The QAZ program then stole additional passwords while providing the intruders with unauthorized access to the computer system.

This scenario illustrates why it's critical for companies to ensure that their employees don't unknowingly provide hackers with access into the corporate network by logging onto corporate computers from unsecured home computers. If organizations are going to allow access to corporate networks from employees at home, there must be appropriate measures and procedures to protect their home computers. As a result, many organizations require that firewalls be installed on their employee's home computers to reduce this vulnerability. Such firewalls are especially critical when a home user is always connected to the Internet via a Digital Subscriber Line (DSL), cable network, or some other permanent connection.

Questions:

1. Some people think that Microsoft downplayed the seriousness of the break-in. What would be the motivation for such a strategy?

2. Why would Microsoft have allowed the attackers access to their network and systems for several days after first detecting the break-in?

3. Imagine that you are in charge of recommending new and improved security measures for Microsoft. What recommendations would you make? How would you justify the costs of implementing your ideas?

Sources: adapted from Steven J. Vaughn-Nichols, "Microsoft Can't Spin This Worm," October 27, 2000, accessed at www.msnbc.com; Stuart Glascock and Mitch Wagner, "Microsoft, FBI, Security Experts Probe Hacking," October 27, 2000, accessed at www.TechWeb.com; Mike Brunker, "Hackers Boldly Break Into Microsoft," October 28, 2000, accessed at www.msnbc.com; Jaikumar Vijayan and Carol Sliwa, "Microsoft

Break-In Points to Security Holes," *Computerworld*, November 6, 2000, accessed at www.computerworld.com; Dan Verton, "Worm Thought to Have Been Used Against Microsoft Had Links to Chinese System," *Computerworld*, January 17, 2001, accessed at www.computerworld.com.

2. Visa Combats Online Credit Card Fraud

Visa-branded credit cards generate almost $2 trillion in annual volume and are accepted at over 22 million locations around the world. Although Visa itself does not offer cards or financial services directly to consumers and merchants, it plays a pivotal role in advancing new payment products and technologies on behalf of its members. Visa also operates a large and sophisticated consumer-payments processing system, known as VisaNet, that can process over 3,700 transactions every second and is capable of handling transactions denominated in 160 different currencies.

Visa transactions from the Internet total roughly $40 billion/year. The company estimates that although just $.06 of every $100 spent with Visa cards as a whole is fraudulent, that number jumps to $.24 for Web transactions. Thus, Visa member financial institutions experience almost $100 million/year in credit card fraud associated with Internet purchases. To combat this problem, Visa worked for over a year with many top Internet solutions providers (including Accenture, Cap Gemini, Ernst & Young, IBM, Microsoft, and Sun Microsystems) to speed the global deployment of Visa Authenticated Payment, a comprehensive e-commerce program designed to ensure safe and secure online payment transactions.

When a cardholder enrolls in Visa Authenticated Payment, the cardholder and his or her financial institution validate the online transaction, just as is done at the check-out stand in the physical world. When a shopper on a PC or a Wireless Application Protocol (WAP)-enabled phone is ready to check out, a pop-up screen is displayed and the customer is required to enter a password after entering his or her credit card information. Before the transaction can be authorized, the password must be authenticated by the bank that issued the customer's Visa card. This new authentication process will be offered to consumers initially on a voluntary basis. It is expected that by 2003, Visa will require all cardholders to establish a password to be used in the authentication process.

Some e-commerce merchants are not convinced that the new authentication process will significantly reduce the incidence of credit card fraud. However, it may help some consumers feel more confident in placing online orders using their credit card. Others argue that people who are frightened of entering their credit card number in an order entry form on the Web won't feel much better about entering their password.

Visa has also implemented the Electronic Compliance Monitoring program, under which participating Internet merchants volunteer to have security experts try to compromise their networks and databases in mock hacking attempts. The friendly attempts at a break-in will ascertain if the e-merchants meet Visa mandated security measures that include designating information security officers, encrypting stored data, and implementing firewalls.

Questions:

1. Imagine that you are the manager of the Visa Authenticated Payment program. How would you measure the success (failure) of the program to determine if it should be continued or stopped?
2. Which program, Visa Authenticated Payment or the Electronic Compliance Monitoring, has the greater potential for reducing credit card fraud? Why?
3. What potential problems can you see with the Visa Authenticated Payment program?

Sources: adapted from "Visa and Internet's Top Industry Players Aggressively Promote Global Adoption of Authenticated Payment," June 27, 2001, accessed at www-s2.visa.com; Maria Atanasov, "The Truth About Internet Fraud," *Smart Business*, April 2001, pp. 92–101; and Dennis Callaghan, "Visa Upgrades Its Web Authentication," *eWeek*, May 14, 2001, pg. 17.

Endnotes

[1] Patrick Thibodeau, "CIS Official Warns Congress of Cyberattack Danger," *Computerworld*, June 21, 2001, accessed at www.computerworld.com.

[2] Stephen H. Wildstrom, "Microsoft Backpedals on the Outlook Security Patch," *Business Week Online*, May 25, 2000, accessed at www.businessweek.com.

[3] Michael Bertin, "The New Security Threats," *Smart Business*, February 2001, pg. 78–86.

[4] Source: Brian Sullivan, "'Cute Trojan Horse' Attacks Through E-mail," *Computerworld*, May 8, 2002, accessed at www.computerworld.com.

[5] Marc L. Songini, "Hospital Confirms Hacker Stole 5,000 Patient Files," *Computerworld*, December 18, 2000, pg. 7.

[6] "What's The Department Of Justice Doing About Identity Theft And Fraud?", accessed at the Department of Justice Web site, at www.usdoj.gov, on May 30, 2002.

[7] Lynn Keith, "Steps to a Secure Network," SANS Institute, May 16, 2001, accessed at www.sans.org.

[8] Weld Royal, "Too Much Trust?" *Industry Week*, November 2, 1998, accessed at www.industryweek.com.

[9] "In-N-Out Gets $256K 'Espionage' Settlement," *Nation's Restaurant News*, April 8, 2002, accessed at www.findarticles.com.

[10] Lee Gomes and Ted Bridis, "FBI Warns of Russian Hackers Stealing Credit-Card Data from U.S. Computers," *The Wall Street Journal*, March 9, 2001, pg. A4.

[11] "Flaw Causes Credit Card Chaos," *MSNBC News*, May 19, 2001, accessed at http://moneycentral.msn.com.

[12] John Schwartz, "Cyberspace Seen As Potential Battleground," *The New York Times on the Web*, November 23, 2001, accessed at www.nytimes.com.

[13] Patrick Thibodeau, "CIS Official Warns Congress of Cyberattack Danger," *Computerworld*, June 21, 2001, accessed at www.computerworld.com.

[14] Patrick Thibodeau, "CIS Official Warns Congress of Cyberattack Danger," *Computerworld*, June 21, 2001, accessed at www.computerworld.com.

[15] "E911 Virus," FBI National Infrastructure Protection Center Web site, at www.nipc.gov, accessed on May 30, 2001.

[16] George V. Hulme and Bob Wallace, "Top of the Week Hacktivism," *Information Week*, November 13, 2000, pg. 22–24.

[17] Patrick Thibodeau, "Federal Agencies Get Poor Grades for Security," *Computerworld*, September 18, 2000, pg. 20.

[18] Patrick Thibodeau, "FAA Faces More Criticism for Computer Security Failings," *Computerworld*, September 27, 2000, accessed at www.computerworld.com.

[19] James Cope, "Faster Firewall Helps Retailer Speed Web Transactions," *Computerworld*, January 7, 2002, accessed at www.computerworld.com.

[20] Alex Salkever, "Inexcusable Insecurity: Microsoft Is Hardly Alone," *Business Week Online*, October 31, 2000, accessed at www.businessweek.com.

[21] "Survey Shows That Insiders Are Main Computer Security Threat," *The New York Times On the Web*, June 19, 2001, accessed at www.nytimes.com.

[22] Pete Loshin, "Intrusion Detection," *Computerworld*, April 16, 2001 acces www.computerworld.com.

[23] Ellen Messmer, "Hackers, Vendors Put Camouflage to Use," *Network World*, February 4, 2002, accessed at www.nwfusion.com.

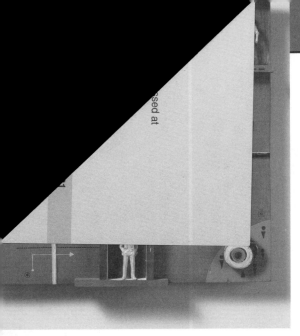

Chapter 4

PRIVACY

"You already have zero privacy—get over it."
— Sun Microsystems CEO Scott McNealy [1]

Qwest Communications is the fourth-largest United States local phone company and also offers Internet, data, multimedia, and voice services to 30 million customers, primarily in the western United States. In January 2002, Qwest found itself embroiled in a public relations controversy over its data privacy policy. In a pamphlet sent to its customers, Qwest stated that unless customers contacted the company to prohibit the practice, it would share data such as telephone services used, billing information, and places called with several subsidiaries and companies with whom it has marketing agreements. Qwest received many complaints from privacy advocates and customers who were concerned about potential invasions of privacy, such as new floods of junk mail and phone calls from travel agents offering fares to places they call. It also received letters from the Electronic Privacy Information Center (EPIC), attorneys general from several states, and a United States senator requesting that Qwest obtain its customers' express consent before sharing information with its business partners.

On January 28, 2002, Qwest announced that it was withdrawing its short-lived plan to share private customer account information. Qwest spokespeople said the firm was listening to its customers who expressed concern and who said they did not understand what Qwest was doing. It also announced the appointment of a chief privacy officer to oversee the review, improvement, and enforcement of the company's privacy policies and procedures.

Sources: adapted from John Schwartz, "Qwest Plan Stirs Protest Over Privacy," *The New York Times on the Web*, January 1, 2002, accessed at www.nytimes.com; "Qwest Shareholders Lash Out at CEO Nacchio, Company," Reuters, June 4, 2002, accessed at http://moneycentral. msn.com; "Qwest Communications Withdraws Plan to Share Private Customer Account Information Within Company," Qwest Press Release, January 28, 2002, found at the EPIC Web site at www.epic.org.

As you read this chapter, consider the following questions:

1. What is the right of privacy and what is the basis for the protection of an individual's privacy under the law?

2. What are the two fundamental forms of data encryption and how does each work?

3. What are the various strategies for consumer profiling and what are the associated privacy issues?

4. What must an organization do to treat consumer data responsibly?

5. Why and how are employers increasingly implementing workplace monitoring?

6. What are the capabilities of the Carnivore system and other advanced surveillance technologies, and what privacy issues are raised by these technologies?

PRIVACY PROTECTION AND THE LAW

The use of information technology in business requires the balancing of the needs of those who use information about individuals against the rights and desires of those individuals whose information may be used.

On the one hand, information about individuals is gathered, stored, analyzed, and reported because organizations can use it to make better decisions. Some of these decisions can profoundly affect the lives of those individuals—hire/not hire a candidate, approve/disapprove a loan, offer/don't offer a scholarship. In addition, the global marketplace and increased competitiveness have caused details about the purchasing habits and financial condition of consumers to become more and more important to companies. They use this information to target their marketing efforts to the consumers most likely to buy their products and services. Organizations also need basic information about their customers to serve them better. It is hard to imagine an organization having any sort of relationship with its customers without having data about them. Thus, organizations seek to implement systems that collect and store key data from every interaction a customer has with an organization.

On the other hand, many individuals feel that information technology designed to meet the information needs of government and business has stripped them of the power to control access to and dissemination of their personal information. The existing hodgepodge of privacy laws and practices fails to provide adequate protection. Instead, it causes confusion that fuels a sense of distrust and skepticism, as illustrated in the Qwest example at the start of the chapter. As a result, consumers often object to the data collection policies of government and businesses. Privacy is a key concern of Internet users and is also a top reason why non-users still avoid the Internet.

A combination of approaches—new laws, technical solutions, and ethical policies—is required to balance the scales. Reasonable limits must be set on government and business access to personal information, new information and communication technologies must be designed in ways that protect rather than diminish privacy, and appropriate corporate policies to set baseline standards for individuals' privacy must be developed. Education and communication are essential as well.

This chapter will provide you with an understanding of the right to privacy and present an overview of a number of information technology developments that have an impact on this right. It will also raise a number of ethical issues that must be addressed by those responsible for gathering data about individuals while continuing to respect the data privacy of consumers, employees, and the public.

First, it is important to gain a historical perspective on the right to privacy. When the United States Constitution was written in 1791, a major concern of the drafters was that a powerful government would intrude on the privacy of individual citizens. As a result, they added the Bill of Rights to protect citizens from too strong a government. So, although the text of the Constitution does not even contain the word privacy, the United States Supreme Court has ruled that the concept of privacy is protected by a number of amendments in the Bill of Rights. For example, the Supreme Court has stated that American citizens have the protection of the Fourth Amendment when there is a "reasonable expectation of privacy." The Fourth Amendment is as follows:

> "The right of the people to be secure in their persons, houses, papers, and effects, against unreasonable searches and seizures, shall not be violated, and no Warrants shall issue, but upon probable cause, supported by Oath or affirmation, and particularly describing the place to be searched, and the persons or things to be seized."

Very importantly though, without a reasonable expectation of privacy, there is no privacy right to protect.

Today, in addition to protection from government intrusions, individuals also need privacy protection from large corporations. There are actually few laws that provide such protection. Most people assume that they have greater privacy rights than the law provides. Some individuals believe that only people with something to hide should be concerned about the loss of privacy; however, we must all be concerned about the potential loss of our right to privacy. As the United States Privacy

Protection Study Commission found in 1977 (when the computer age was in its infancy), "The real danger is the gradual erosion of individual liberties through the automation, integration, and interconnection of many small, separate record-keeping systems, each of which alone may seem innocuous, even benevolent, and wholly justifiable."

The Right of Privacy

A broad definition of the right of privacy is that it is "the right to be left alone—the most comprehensive of rights, and the right most valued by a free people" (Justice Louis Brandeis dissenting in *Olmstead v. U.S., 1928*). Another definition, particularly useful in discussing the impact of information technology on individual rights, is that it is "the right of individuals to control the collection and use of information about themselves."[2]

As a legal concept, the right to privacy has four aspects: protection from unreasonable intrusion upon one's isolation (such as the gathering of details about an individual's Web surfing habits), protection from appropriation of one's name or likeness (such as identity theft that involves stealing one's credit cards or social security number), protection from unreasonable publicity given to one's private life (such as the revealing of details about one's medical condition), and protection from publicity that unreasonably places one in a false light before the public (such as the publishing of false information about an individual on a Web site).[3] As shown by the preceding examples, the use of information technology can lead to a potential violation of all four aspects of the right to privacy.

Recent History of Privacy Protection

This section outlines a number of legislative actions taken over the past 40 years that affect an individual's privacy. Note that most of these actions address invasion of privacy by the government. Legislation protecting individuals from data privacy abuses by private industry are almost nonexistent. In addition, although a number of independent laws and acts have been implemented over time, no single overarching national data privacy policy has been developed. As a result, existing legislation is sometimes inconsistent and even conflicting at times.

The Communications Act of 1934 restricted the government's ability to secretly intercept communications. However, under a 1968 Federal statute, wiretapping (the interception of telephone or telegraph communications for purpose of espionage or surveillance) can be done by law enforcement officers, provided a court order has first been obtained.

The Freedom of Information Act (FOIA) was passed in 1966 and amended in 1974. It allows the public better access to government records to find out what the government knows about them and what policies agencies use to govern the public. Read the Legal Overview to find out more about the FOIA.

FREEDOM OF INFORMATION ACT (FOIA)

The Freedom of Information Act (FOIA), passed in 1966 and amended in 1974, provides the public with the means to gain access to certain government records. The public can use well-defined FOIA procedures to find out the spending patterns of an agency, the agency's policies and the reasoning behind them, and the agency's mission and goals. FOIA is often used by whistleblowers to obtain records that they otherwise would not be able to obtain. Citizens have also used the FOIA to find out what information the government has about them.

The first part of the FOIA outlines the kinds of information that all government agencies are required to publish in the *Federal Register*, which is a compilation of government news and data. The second part of the law outlines the process to enable the public to view the records of agencies. The FOIA has clearly defined procedures to obtain information. The process can be summarized as follows: You must submit your request for records in writing to the appropriate agency. You must state specifically what information you are requesting and the purpose of the information, such as commercial use, private-information use, scientific or educational use, or mass media use. Each agency has an identified freedom of information officer who processes all requests. By law, the agency must respond to your request within ten days; an additional ten days is allowed due to a work backlog or difficulties in finding the information you are requesting. If your initial request is denied, you may file a FOIA appeal. Agencies have 20 days to respond to your appeal or you can take your claim to federal district court.

Exemptions bar disclosure of information that could compromise national security or interfere with an active law enforcement investigation. Another exemption prevents disclosure of records if that disclosure would cause an invasion of someone's privacy. In this case, a balancing test is applied to evaluate if the privacy interests at stake are outweighed by competing public interests.

Some legislators are calling for an additional exemption to protect information shared between the government and private industry with regard to information about attacks against computer and information systems. The sharing of such information is important for the federal government to protect the nation's critical information technology infrastructure from attack. However, private companies are very reluctant to share such data. They fear that the FOIA doesn't offer enough assurances that their proprietary data will remain secret. A survey by the San Francisco-based Computer Security Institute and the FBI shows that in 2000, only 25 percent of companies who had experienced hacks and intrusions reported them to law enforcement agencies. Of those that didn't report the incidents, 52 percent said they feared negative publicity and 39 percent cited concerns about competitors using the information to their advantage. Critics of the information-sharing legislation say private industry has a faulty perception of the law and that further exemptions would only serve to weaken the FOIA.

Agencies can charge for the costs associated with fulfilling your FOIA request, but they must tell you how much they will charge before processing your request. Usually public information can be obtained for a charge that covers the incremental cost of providing it—the cost of a CD-ROM and the postage to mail it to the requester or a few cents per minute for connect time.

Government agencies try to balance the individual's right to privacy with the demands placed on them by the FOIA. Controversies often break out when an agency decides to put public information online. For example, the county of Hamilton, Ohio, created and published on the Web a real estate database showing the owner and purchase price of every home in the county. A potential buyer or nosy neighbor can easily look up an address and see what the owner paid for the home, its current value, and when it was purchased. The county auditor argued that the data was a matter of public record and had always been available for anyone requesting it. Making the database accessible via the Internet reduced the auditor's workload. On the other hand, many homeowners do not want their neighbors or prospective buyers to have such quick and easy access to the purchase prices of their homes.

Sources: adapted from Dan Verton, "Rule Changes May Further Protect Company Security Data," *Computerworld*, July 9, 2001, accessed at www.computerworld.com, Eric J. Sinrod, "Does Sharing Violate Security or Help Preserve It?", *Computerworld*, May 30, 2001, accessed at www.computerworld.com; "Freedom of Information Act," accessed at the Electronic Privacy Information Center at *www.epic.org* on July 13, 2002.

The Fair Credit Reporting Act of 1970 regulates the operations of credit-reporting bureaus, including how they collect, store, and use credit information. It is designed to promote accuracy, fairness, and privacy of information in the files of credit reporting companies (such as Experian, Equifax, and Trans Union) and to check verification systems that gather and sell information about individuals. The act outlines specific rules that govern things such as who may access your credit information, how you can find out what is in your file, how to dispute inaccurate data, and how long data is retained. The manual procedures and information systems of the credit reporting bureaus must implement and support all these regulations.

The Privacy Act of 1974 limits how the United States government collects, maintains, uses, and disseminates personal information. Its purpose is to provide certain safeguards for individuals against invasion of personal privacy by federal agencies. The Central Intelligence Agency (CIA) and law enforcement agencies are excluded from this act. This act does not cover the actions of private industry.

The Organization for Economic Cooperation and Development (OECD) is an international organization consisting of 30 member countries (including Australia, Canada, France, Germany, Italy, Japan, Mexico, New Zealand, United Kingdom, and the United States). Its goal is to set policies and make agreements in areas where multilateral agreement is necessary for individual countries to make progress in a global economy. Dialogue, consensus, and peer pressure are essential to make these policies and agreements "stick."[4] The 1980 privacy guidelines set by the OECD—also known as the Fair Information Practices—are often held up as a model of ethical treatment of consumer data for organizations to adopt. These guidelines are composed of eight principles: collection limitation, data quality, purpose specification, use limitation, openness, security safeguards, individual participation, and accountability. These principles are summarized in Table 4-1.

Table 4-1   Summary of the 1980 OECD privacy guidelines

Principle	Guideline
Collection Limitation	Limit the collection of personal data. All such data must be obtained by lawful and fair means with the subject's consent and knowledge.
Data Quality	Personal data should be accurate, complete, current, and relevant to the purpose for which they are used.
Purpose Specification	The purpose for which personal data is collected should be specified and that purpose should not be changed.
Use Limitation	Personal data should not be used for other than the specified purpose without the consent of the individual or by authority of law.
Security safeguards	Personal data should be protected against unauthorized access, modification, or disclosure.
Openness principle	Data policies should exist and a "data controller" should be identified.
Individual Participation	An individual should have the right to review his/her data, to challenge the correctness of the data, and to have incorrect data changed.
Accountability	A "data controller" should be responsible for ensuring that the above principles are met.

Source: "OECD Guidelines on the Protection of Privacy and Transborder Flows of Personal Data" pages 14–18, ©2002.

The Electronic Communications Privacy Act (ECPA) of 1986 amended Title III of the Omnibus Crime Control and Safe Streets Act of 1968, extending Title III's existing prohibitions against the unauthorized interception of electronic communications. Title III and the ECPA together prohibit unauthorized interception, disclosure, or use of one's oral or electronic communications. However, the ECPA failed to address emerging technologies such as wireless modems, radio-based electronic mail, and cellular data networks; thus these communication technologies can still be legally intercepted.

In 1994, Congress adopted the Communications Assistance for Law Enforcement Act (CALEA), requiring telephone companies to design their systems to ensure a certain basic level of government access. CALEA is highly controversial, and many argue that it enables the FBI to greatly expand its eavesdropping capabilities. In its final form, it requires telecommunications carriers to alter their voice networking technology so the FBI can conduct court-ordered wiretaps without being hampered by new digital services such as call-forwarding. The measure also budgeted $500 million in federal funding over four years to pay the telecommunication companies for the necessary technical changes.

In June 1998, a Federal Trade Commission (FTC) report showed that 85 percent of Web sites collected personal information about consumers, while only 14 percent of those companies provided any notice to Web surfers that such information gathering was taking place.[5] The same report pointed out that children were especially at risk. The FTC recommended congressional action and, acting with uncharacteristic haste, Congress passed the Children's Online Privacy Protection Act (COPA) within four months. According to the COPA law, Web sites that cater to children must offer comprehensive privacy policies on their sites, notify their parents or guardians about their data collection practices, and receive parental consent before they collect any personal information from children under 13 years of age. The law went into effect in April 2000 and has made a major impact—while many companies are spending hundreds of thousands of dollars to make their sites compliant, others are eliminating preteens as a target audience.[6] (In April 2001, three Web sites were fined a total of $100,000 for collecting personal information from children without getting permission from their parents. The FTC also alleged that none of the three Web sites posted privacy policies as required by COPA).[7] Privacy advocate groups including the Electronic Frontier Foundation, the American Civil Liberties Union (ACLU), and the Electronic Privacy Information Center filed suit, claiming that the law violates First Amendment rights to free speech. Since then, a string of injunctions have kept it from being fully enacted. Although a federal appeals court declared that COPA stymies online speech, the United States Department of Justice (DOJ) appealed the ruling to the Supreme Court in 2002 in hopes of getting the law reinstated.[8] Although the DOJ argues that COPA balances adults' free speech rights with the protection of children online, opponents claim that the language is overly vague. In May 2002, the Supreme Court issued a decision on the COPA. Although the Supreme Court did not decide any of the core legal questions, it ordered a lower court to decide the case on a wider range of First Amendment issues. It also left in place an injunction barring enforcement of the law.

The European Community Directive 95/46/EC of 1998 requires that any company doing business within the borders of 15 Western European nations put in place a set of privacy directives on fair and appropriate use of information.[9] Basically, this directive says that member countries must ensure that data transferred to non-European Union (EU) countries is protected. It also bars the export of data to countries that do not have data privacy protection standards comparable to the EU. Initially, the EU countries were concerned that the largely voluntary system of data privacy of the United States did not meet the EU directive's stringent standards. Eventually, the United States Department of Commerce worked out a safe harbor agreement that would allow American companies to import and export personal data.[10] However, addressing this issue has clearly distinguished the European philosophy of strict government regulation with enforcement by a set of commissioners from the United States philosophy of having no federal privacy policy and with self-regulation overseen by the United States Department of Commerce, the Better Business Bureau Online (BBB Online), and TRUSTe.

The Better Business Bureau Online and TRUSTe are independent, non-profit privacy initiatives that favor an industry-regulated approach to data privacy. They are concerned that strict government regulation could have a negative impact on the Internet's use and growth and that such regulation would be costly to implement and difficult to change or repeal. A Web site operator can apply for either the BBB Online or TRUSTe data privacy seal in order to demonstrate that a site adheres to a high level of data privacy. Having one of the seals can increase consumers' confidence in the site operator's desire and ability to manage data responsibly. Thus, the presence or absence of the BBB Online or TRUSTe seal enables users to make more informed decisions about whether to release personal information (such as phone numbers, addresses, and credit card numbers) to a Web site.

An organization must join the Better Business Bureau and pay a small annual fee ranging from $200 to $7,000, depending on annual sales, before applying for the BBB Online Privacy seal. Because the BBB Online Privacy seal program identifies online businesses that honor their own stated privacy protection policies, the Web site operator must also have adopted and posted an online privacy notice. For a Web site to receive the TRUSTe seal, its operators must demonstrate that it adheres to established privacy principles. They must also agree to comply with TRUSTe's oversight and consumer resolution process. There is also an annual fee that must be paid. The privacy principles included require that the Web site openly share what personal information is being gathered, how it will be used, with whom it will be shared, and whether the user has an option to control its dissemination.

An FTC report in 1999 noted improvement in Web sites notifying their users about the collection of personal information, but at only a 66 percent compliance rate.[11] The FTC, privacy advocates, and consumers expressed concern over industry self-regulation. Within a few months, the FTC provided recommended legislation to Congress that included baseline privacy regulations. In addition, various members of Congress have proposed several competing privacy-related bills. A key issue for these bills is the policy of opt-out or opt-in for information gathering. **Opt-out** assumes that consumers approve of companies collecting and storing their personal information, and it requires them to take action to opt-out by specifically telling companies, one by one, not to collect data about them. **Opt-in** requires the data collector to get specific permission from a consumer before collecting any of his or her data.[12] Data collectors favor opt-out, and consumer groups favor opt-in.

One example of a bill legislating opt-out information gathering is the 1999 Gramm-Leach-Biley Act, which required all financial-services institutions to communicate their data privacy policies and honor customer data gathering preferences by July 1, 2001. Its goal was to cause all financial institutions to take appropriate actions to protect and secure customer's nonpublic data from unauthorized access or use. To comply, institutions resorted to mass mail to contact their customers with privacy-disclosure forms. As a result, many people received a dozen or more similar-looking forms—one from each financial institution with which they do business. However, most of the forms looked "like car warranties. They were densely worded, packed with small-type legalese, and were forbiddingly designed."[13] Rather than making it easy for customers to opt-out, the documents required that consumers send one of their own envelopes to a specific address and state in writing that they wanted to opt-out—all this rather than a simple turnaround postcard allowing the customer to check off his or her choice. As a result, most customers threw out the forms without grasping the full implications of what was being asked. These customers thus "opted in" and gave the financial institutions the right to sell information such as annual earnings, net worth, employers, specific investments, loan amounts, and social security numbers to other financial institutions.

The 2001 Uniting and Strengthening America by Providing Appropriate Tools Required to Intercept and Obstruct Terrorism (USA Patriot Act) was passed in response to the September 11, 2001, terrorism acts. It gives sweeping new powers to both domestic law enforcement and international intelligence agencies. The act is large (it is over 342 pages long) and complex (it changes over 15 existing statutes), but it took just five weeks from its first introduction until it was passed into law. Legislators rushed to get the act approved in the house and senate, arguing that law enforcement authorities needed these powers to help track down terrorists and prevent future attacks. Critics argue that the law removes many of the checks and balances that previously gave the courts the opportunity to ensure that law enforcement agencies did not abuse their powers. They also argue that many of its provisions have nothing to do with fighting terrorism. Some of the more controversial sections of the act include the following:

- The crime of cyberterrorism is defined as hacking attempts that cause $5,000 in aggregate damage in one year, damage to medical equipment, or injury to any person. Those convicted of cyberterrorism are subject to a prison term of five to twenty years.
- Internet service providers (ISPs) and telephone companies must turn over customer information, including numbers called, without a court order, if the FBI claims that the records are relevant to a terrorism investigation. Furthermore, the company is forbidden to disclose that the FBI is conducting an investigation.
- The FBI is granted broad access to sensitive medical, financial, mental health, and educational records without requiring evidence of a crime and without a court order.
- Federal agents can subpoena customer payment records to obtain the identity of a user behind an e-mail address.
- Any federal judge, regardless of jurisdiction, can now approve warrants for law enforcement officers to bypass Internet providers to read e-mail and wiretap conversations in real time.

In June 2002, Attorney General John Ashcroft proposed Justice Department regulations requiring some 100,000 foreign visitors, including students, tourists, and researchers, to register with the federal government, to get fingerprinted and photographed, and to fill out a long personal information form. Mr. Ashcroft stated that the rules were an important beginning toward meeting a congressionally mandated goal to implement an information system by 2005 to track all of the 35 million annual foreigners who visit the United States. The outcome of Mr. Ashcroft's proposal is undecided as of this writing. Some members of Congress see the rules as a means to regain control over illegal immigration in the United States, while others view them as troubling and poorly thought out.[14]

The United States has neither enacted the legislation to establish minimal privacy standards nor established an advisory agency that could recommend acceptable privacy practices to businesses. Instead, there are laws that address potential abuses by the government with little or no restrictions for private industry. You can track the status of current legislature aimed at privacy at the Electronic Privacy Information Center's site at *www.epic.org*.

KEY PRIVACY AND ANONYMITY ISSUES

This section discusses a number of current and important privacy issues including data encryption, customer profiling, the need to treat customer data responsibly, workplace monitoring, spamming, the FBI's Carnivore system, and advanced surveillance techniques.

Data Encryption

Cryptography is the science of encoding messages so that only the sender and the intended receiver can understand them. Cryptography was once a mysterious science employed mainly during wartime to make it impossible for enemies to understand each others' communications. For example, during World War II, the United States sent radio messages between troops spoken in the native language of the Navajo Indians, a "code" that the Axis powers were unable to crack. Today cryptography is a key tool for ensuring the confidentiality, integrity, and authenticity of electronic messages and business transactions sent between computer users and information systems.

Encryption is the process of converting an original electronic message into a form that can be understood only by the intended recipients. In cryptography, a **key** is a variable value that is applied (using an algorithm) to a string or block of unencrypted text to produce encrypted text or to decrypt encrypted text. The length of the key used to encode and decode messages determines the strength of the encryption algorithm. Encryption methods rely on the limitations of computing power for their security—if breaking a code requires too much computing power, even the most determined code crackers will not be successful.

A **public key encryption system** uses two keys to encode and decode messages. One key of the pair, the message receiver's public key, is readily available to the public and is used by anyone to send that individual encrypted messages. The second key, the message receiver's private key, is kept secret and is known only by the message receiver. Its owner uses the private key to decrypt messages—convert an encoded message back into the original message. For example, if Company A wishes to send an encrypted message to Company B, it must first obtain Company B's public key from what is known as a certifying authority, a kind of Department of Motor Vehicles for encryption users. These certifying authorities maintain a list of companies and their public keys. Then, Company A encrypts the message to Company B using Company B's public key. Once the message is encrypted, only Company B can read the message by decrypting it using its private key. The public key-private key combinations are unique, so only one private key can open a message encrypted with its corresponding public key. Obviously, a company must keep its private key well protected to ensure the security of encrypted messages. An analogy to public key encryption is the use of two keys, one that locks your front door, and one that can unlock it. You could keep a "public key" under your doormat that can be used to lock your front door, but you would keep possession of the "private key" that can unlock it. Knowing an individual's public key does not enable you to decrypt an encoded message to that individual. RSA is a public key encryption algorithm that has been available since 1978. Named for its inventors, Ronald Rivest, Adi Shamir, and Leonard Adleman, RSA is the basis on which much of the security protecting consumers and merchants on the Web has been built.[15] Pretty Good Privacy (PGP) is a software encryption product that uses 128-bit encryption and is the de facto standard for Internet e-mail encryption.

A **private key encryption system** uses a single key to both encode and decode messages. Obviously, both the sender and receiver must know the key to communicate. It is critical that no one else learns the key or else all messages between the two can be decoded by others. Indeed, the issue of distributing the private key without others learning its value is a problem in such systems. An analogy to private key encryption is the use of a single key that both locks and unlocks your front door. Whoever is in possession of the key can gain access to your home. The Digital Encryption Standard (DES), the standard for commercial private key encryption for over 20 years beginning in 1977, employs a 56-bit key and requires anyone trying to crack the code to try over 7.2×10^{16} different combinations.[16] Modern computers were able to crack messages sent using the DES code in less than three days in July 1998. As a result, Triple Data Encryption Standard (3DES), which uses 112- or 168-bit keys, replaced the aging DES.

After a three-year worldwide competition, the National Institute of Standards and Technology (NIST, part of the U.S. Department of Commerce) announced in October 2000 that an encryption algorithm from Belgium would become the Advanced Encryption Standard (AES) for the United States, replacing DES. The new encryption algorithm is called Rijindael and was developed by Belgian cryptographers Joan Daemen and Vincent Rijmen. This algorithm would require code crackers to try as many as 1.1×10^{77} combinations and would take current technology computers over 149 trillion years of computing time to decode a message encrypted using this algorithm.[17] However, future advances in computer technology can be expected to render even this code "crackable," but at least not for a decade or two.

The United States government places export limitations on hardware and software that employ data encryption. This is done to prevent such technology from falling into the hands of its enemies. If found guilty of violating the Arms Export Control Act, individuals can be jailed for ten years and fined $1 million. Once extremely restrictive, these limitations have been loosened to increase exports. United States companies are able to export encryption products to customers in the European Union, Australia, Czech Republic, Hungary, Japan, New Zealand, Norway, Poland, and Switzerland. In spite of these restrictions, intelligence agencies say that terrorist groups such as Osama bin Laden's al Qaeda network have used encryption to protect their phone messages and e-mail.

Consumer Profiling

Companies openly collect personal information about people who surf the Internet when those people register at sites, complete surveys, fill out forms, or enter contests online. Many companies also obtain information about Web surfers without their manual input, through the use of **cookies** (a text file that a Web site puts on your hard drive so that it can remember something about you at a later time) and tracking software that allow companies to use their Web sites to analyze browsing habits to deduce personal interests and preferences. The use of cookies and tracking software is controversial because it enables companies to find out information about consumers without their explicit permission. Outside of the Web environment, marketing firms employ similarly controversial means to collect information about people and their buying habits. Each time a consumer uses a credit card, redeems frequent flyer points, fills out a warranty card, answers a phone survey, buys groceries using a store loyalty card, orders from a mail-order catalog, or registers a car with the DMV, data is added to a storehouse of personal information about that consumer. In none of these cases does the consumer explicitly consent to submitting his or her information to a marketing organization.

Marketing firms aggregate the information they gather about consumers from disparate sources to build databases containing a huge amount of consumer behavioral data. They want to know as much as they can about consumers—who they are, what they like, how they behave, and what motivates them to buy. The marketing firms provide this data to companies to enable them to tailor their products and services to individual consumer preferences. Advertisers use the data to more effectively target and attract customers to their messages. Ideally, this means that buyers should be able to find or receive product offers well-suited for them and experience a more efficient buying process. Sellers should be better able to tailor their products and services to meet their customers' desires and to increase sales. However, concerns about how all this data actually gets used is the single leading cause of hesitation for those potential Web shoppers who have yet to make online purchases.[18]

Large-scale marketing organizations such as DoubleClick employ advertising networks to serve ads to thousands of Web sites. When someone clicks on an ad at one company's Web site, tracking information about that person, from what he or she clicked to what he or she bought, is gathered and forwarded to DoubleClick, where it is stored in a large database. A collection of Web sites served by a single advertising network is called a collection of **affiliated Web sites**.

Marketers use cookies to recognize return visitors to their sites and to store useful information about them. The goal is to provide customized service tailored for each consumer. When someone visits a specific Web site, the site "asks" that person's computer if it can store a cookie on the hard drive. If the computer agrees, it is assigned a unique identifier and a cookie with this identification number is placed on its hard drive. During a Web-surfing session, three types of data are gathered. First, as one browses the Web, "GET" data is collected. **GET data** is data that reveals, for example, that the consumer visited an affiliated book site and requested information about the latest Dean Koontz book. Second, "POST" data is captured. **POST data** is the data entered into blank fields on an affiliated Web page when a consumer signs up for a service, such as the Travelocity service that sends an e-mail when airplane fares change for flights to favorite destinations. Third, the marketer monitors the consumer's surfing throughout any affiliated Web sites, keeping track of the information the user sought and viewed. This is known as **click-stream data**. Thus as a single person surfs the Web, a tremendous amount of data of interest to marketers and sellers is generated.

After a cookie has been stored on a computer, the next time someone using that computer visits a site (or any affiliated site in the case of a network advertiser), the browser is asked to check if it has stored a cookie on your hard drive. If so, the cookie is then used to search for information about you in the network advertiser's database. If a match is found, whatever information might be stored there about the individual can be used to tailor the ads and promotions that are presented to him or her. The marketer also knows what ads have been seen most recently and makes sure that they aren't seen again (unless the advertiser has decided to market using repetition). The marketer also tracks what sites are visited and uses that data to make educated guesses about the kinds of ads that would be of most interest to the user.

It is possible to limit or even stop the deposit of cookies on a hard drive. There are four ways to do this: Browsers can be set to limit or stop cookies; cookies can be manually deleted from a hard drive; one of the many cookie-management programs available for free on the Web can be installed; or consumers can use anonymous browsing programs that don't accept cookies. However, an increasing number of Web sites lock visitors out unless they allow cookies to be deposited on their hard drives.

Other organizations besides marketers use cookies. In 2000, it was uncovered that cookies were planted on the hard drives of visitors to the Web site of the White House's Office of National Drug Control Policy. They admitted the cookies placed on visitors' computers could be used to find leads to people who might be growing marijuana or selling cocaine.[19]

In addition to utilizing cookies to track consumer data, **personalization software** is used by marketers to optimize the number, frequency, and mixture of their ad placements. It also is used to evaluate how visitors react to new ads. The goal is to turn first-time visitors to a site into paying customers and to facilitate greater cross-selling activities.

There are several types of personalization software. **Rules-based** personalization software uses business rules tied to customer-provided preference information or online behavior to determine the most appropriate page views and product information to display. For instance, if you use a Web site to book airline tickets to a popular vacation spot, rules-based software might ensure that you are shown ads for rental cars. **Collaborative filtering** offers consumer recommendations based on the types of products purchased by other individuals who bought the same product as another customer. For example, if you bought a book by Dean Koontz, a company might recommend Stephen King books to you, based on the fact that a significant percentage of other customers have bought books by both authors. **Demographic filtering** is another form of personalization software that augments click-stream data and user-supplied data with demographic information associated with user Zip codes to make product suggestions. Yet another form of personalization software, **contextual commerce**, associates product promotions and other e-commerce offerings with specific content a user may be receiving in a news story online. For example, as you read a story about white-water rafting, you may be offered deals to buy rafting gear or a promotion for a vacation in West Virginia, a state with many rivers suited for white-water rafting.[20] Instead of simply bombarding customers at every turn with standard sales promotions that get tiny response rates, marketers are getting smarter about where and how they use personalization. They are also taking great care to measure whether personalization is paying off. The intended result is that effective personalization increases online sales and improves the consumer relationship.

Online marketers cannot capture personal identification information, such as names, addresses, and social security numbers, unless individuals provide them. Without this information, companies can't separately contact individual Web surfers who visit their Web sites.[21] Data gathered about a Web surfer's Web browsing through the use of cookies is anonymous as long as the network advertiser doesn't link that data with personal identification information. However, if one volunteers that information while visiting a Web site, a Web site operator can use this information to find out additional information about the person that the person may not want the site operator to have. For example, a name and address can be used to find a corresponding phone number, which might then lead to obtaining even more personal data. All this information becomes extremely valuable to the Web site operator who is trying to build a relationship with the person and turn him or her into a customer. The operator can use this data to initiate contact or sell the information to other organizations with which they have marketing agreements.

Consumer data privacy has grown into a major marketing issue. Companies that can't protect or don't respect their customer's information will suffer through loss of customers and potential class-action lawsuits stemming from privacy violations.[22] For example, privacy groups spoke out vigorously to protest the proposed merger of Web ad server DoubleClick and consumer database Abacus. Their concern was that the information stored in cookies would be combined with data from mailing lists, thus revealing the Web users' identities. This would enable the network advertiser to identify and track the habits of unsuspecting consumers. Public outrage and the threat of lawsuits forced DoubleClick to back off this plan.[23]

Opponents of consumer profiling are also concerned that the data being gathered is sold to other companies without the permission of the consumer providing the data. After the data has been collected, consumers have no way of knowing how it is being used or who is using it. For example, when Toysmart.com went bankrupt in 2000, it planned to sell the customer information from its Web site to the highest bidder in order to earn cash to pay its employees and creditors. This data included the names, addresses, and ages of customers and their children. TRUSTe had licensed Toysmart.com to put the TRUSTe privacy seal on its Web site, provided that Toysmart.com never divulged customer information to a third party. Because Toysmart.com was planning to violate that agreement, TRUSTe submitted a legal brief asking the bankruptcy court to withhold its approval for the proposed sale. TRUSTe officials also registered a complaint with the FTC, who launched an investigation and then filed suit to stop Toysmart.com from selling its customer list and related information in violation of the privacy policy that appeared on the company's Web site. Finally, Walt Disney Company, which owned 60 percent of Toysmart.com, bought the list and "retired it" to protect customers' privacy and put an end to the controversy.[24]

One potential solution to consumer privacy concerns is a screening technology called the **Platform for Privacy Preferences (P3P)**, which shields users from sites that don't provide the level of privacy protection they desire. Instead of forcing users to find and read through the privacy policy for each site they visit, P3P software in a computer's browser will download the privacy policy from each site, scan it, and notify the user if the policy does not match his or her preferences. (Of course, unethical marketers can post a privacy policy that does not accurately reflect the manner in which data is treated.) The World Wide Web Consortium, an international industry group whose members include Apple, Commerce One, Ericsson, and Microsoft, initiated and is supporting the development of P3P.

Treating Consumer Data Responsibly

When dealing with consumer data, strong measures are required to avoid the development of customer relationship problems. The most widely accepted approach to treating consumer data responsibly is for a company to adopt the Code of Fair Information Practices and the 1980 Organization for Economic Cooperation and Development (OECD) privacy guidelines. Under these guidelines, an organization collects only the personal information that is necessary to deliver its product or service. The company ensures that the information is carefully protected and that it is accessible only by those with a need to know, and it provides a process for consumers to review their own data and make corrections. The company informs customers if it intends to use customer information for research or marketing and provide a means for them to opt-out.[25]

An increasing number of companies are also appointing executives to oversee their data privacy policies and initiatives. As a result of the increased focus on data privacy, companies recognize the need to establish corporate data privacy policies. Some companies are appointing a **chief privacy officer (CPO)**, while others are assigning these duties to other senior managers. An effective CPO can avoid violating government regulations and reassure customers that their privacy will be protected. This requires that the organization instill in the CPO the power to stop and/or modify major company initiatives. The CPO's general duties include training employees about privacy, checking the company's privacy policies for potential risks, figuring out if gaps exist and how to fill them, and developing and managing a customer privacy dispute and verification process.[26]

The CPO should be briefed on *planned* marketing programs, information systems, or databases that involve the collection or dissemination of consumer data. The rationale for early involvement in such initiatives is to ensure that potential problems can be identified in the earliest stages of the project when it is easier and cheaper to fix them. Some organizations fail to address privacy issues early on, and it takes a negative experience to cause them to appoint a CPO. For example, United States Bancorp, a bank with over $86 billion in assets, appointed a CPO in August 2000, but only after spending $3 million to settle a lawsuit that accused the bank of selling confidential customer financial information to telemarketers.[27] Table 4-2 provides useful guidance for ensuring that your organization treats consumer data responsibly. The preferred answer to each question is "Yes."

Table 4-2	Manager's checklist for treating consumer data responsibly

Questions	Yes	No
Do you have a written data privacy policy that is followed?	____	____
Can consumers easily view your data privacy policy?	____	____
Are consumers given an opportunity to "opt-in" or "opt-out" of your data policy?	____	____
Do you collect only the personal information that is necessary to deliver your product or service?	____	____
Do you ensure that the information is carefully protected and that it is accessible only by those with a need to know?	____	____
Do you provide a process for consumers to review their own data and make corrections?	____	____
Do you inform your customers if you intend to use customer information for research or marketing and provide a means for them to opt-out?	____	____
Have you identified an individual who has full responsibility for implementing your data policy and dealing with consumer data issues?	____	____

Workplace Monitoring

As discussed in Chapter 2, many organizations have developed a policy on the use of information technology to protect against employee abuses that reduce worker productivity or that could expose the employer to harassment lawsuits. The institution and communication of such an IT usage policy establishes the boundaries of acceptable and unacceptable behavior and enables management to take action against those who violate the policy.

The potential for decreased productivity, coupled with increased legal liabilities from computer users, have forced employers to monitor workers to ensure that the corporate IT usage policy is followed. Currently, 78 percent of major United States firms find it necessary to record and review employee communications and activities on the job, including phone calls, e-mail, Internet connections, and computer files.[28] Some are even videotaping employees on the job. In addition, some companies employ random drug testing and psychological testing. With few exceptions, these increasingly common, and many would say intrusive, practices are perfectly legal.

The Fourth Amendment of the Constitution protects citizens from unreasonable searches by the government and is often used to protect the privacy of government employees. Public sector workers can appeal directly to the "reasonable expectation of privacy" standard established by the Supreme Court ruling in *Katz v. United States* (Rodriguez 1998).[29]

However, the Fourth Amendment cannot be used to limit the manner in which a private employer treats its employees, because such actions do not constitute government actions. As a result, public sector employees have far greater privacy rights than those working in private industry. Although private sector employees can seek legal protection against an invasive employer through claims brought under various state statutes, the degree of protection provided by these laws varies widely from state to state. Furthermore, state privacy statutes tend to favor employers over employees. For example, to sue successfully, employees must prove that they have been operating in a work environment where they had a reasonable expectation of privacy. As a result, courts typically rule against employees who file privacy claims for being monitored while using company equipment. It is easy for a company to defeat a privacy claim simply by proving that an employee had been given explicit notice that e-mail, Internet use, and files on company computers were not private and that their use by the employee might be monitored. When an employer engages in workplace monitoring, though, it must ensure that it does not treat one class of worker differently than another for violations of corporate policy. For example, a company could get into legal trouble for punishing an hourly-paid employee more seriously for visiting inappropriate Web sites than it punished a monthly-paid employee.

Our society is struggling to define the extent to which employers should be able to monitor the work-related activities of employees. On the one hand, employers must be able to guarantee a work environment conducive to all workers, ensure a high level of worker productivity, and avoid the costs of defending against "nuisance" lawsuits. On the other hand, privacy advocates want federal legislation that keeps employers from infringing upon the privacy rights of employees. Such legislation would require mandatory, prior notification to all employees of the existence and location of all electronic monitoring devices. They also want to see restrictions placed on the types of information collected and the extent to which an employer may utilize electronic monitoring. As a result of all this, there are many laws being introduced and debated at both the state and federal level. As the laws governing employee privacy and monitoring continue to evolve, business managers must keep informed to avoid having to enforce outdated usage policies. Organizations with global operations face an even bigger challenge, because the legislative bodies of other countries also debate these issues.

Spamming

Spamming is the sending of copies of the same e-mail message to a large number of people in an attempt to cause those people to read a message they might otherwise choose not to receive. Most spam is commercial advertising, sometimes for questionable products such as hard-core pornography and phony get-rich-quick schemes. Spammers frequently send the same message to dozens of Usenet newsgroups (some newsgroups contain over 1,000 members). Spammers also target individual users with direct e-mail messages, building their mailing list by scanning Usenet postings, buying other's mailing lists, or searching the Web for addresses.

Spam is actually an extremely inexpensive method of marketing and is used by many legitimate organizations. For example, some companies will send e-mail to a broad cross-section of potential customers announcing the release of a new product in an attempt to increase initial sales of the product. The cost of creating an e-mail campaign for a product or service is on the order of one thousand dollars compared to tens of thousands of dollars for direct-mail campaigns. In addition, e-mail campaigns take three weeks to develop compared to three months for direct mail, and the turnaround time for feedback averages 48 hours for e-mail, as opposed to three weeks for direct mail.[30] However, the benefits of spam to companies can be largely offset by the generally negative perception the public has of receiving unsolicited ads.

Spam forces sometimes unwanted and objectionable materials into e-mail boxes, detracts from Internet users' ability to communicate effectively (because of full mailboxes and relevant e-mails hidden among many unsolicited messages), and costs Internet users and service providers millions of dollars annually. Usenet members at times can be overwhelmed with a barrage of spam and become frustrated that their Usenet group seems to have lost focus. It takes users time to scan and delete e-mail spams and, for those who pay for Internet connect charges on an hourly basis, this cost can add up.

It also costs money for ISPs and online services to transmit spam, and these costs are reflected in the rates charged subscribers. In early 2001, AOL/Time Warner sued the operators of a Web-based pornography company for allegedly sending millions of unsolicited e-mails with sexual content to its subscribers. AOL/Time Warner also alleged that the messages were sent haphazardly with no attempt to target just those members who had expressed an interest in viewing adult material. Furthermore, many messages ended up in member accounts accessible by children. As one might expect, such unsolicited messages angered subscribers and resulted in numerous complaints—as many as 250,000 in a single day. AOL/Time Warner sought $10 in damages for each unsolicited e-mail message plus $25,000 for each day during which such messages were transmitted via AOL/Time Warner's computer network.[31]

To avoid costs, unethical spammers frequently use fake return addresses on their messages so that they don't have to pay for receiving responses from people to whom they've sent messages. They also frequently use "disposable" trial ISP accounts (that they let expire) so that the ISP bears the cost of cleaning up after them.

Many spam messages request that the recipient "please send a REMOVE message to get off our list"—which raises the question of why one should have to do anything to get off a list one never asked to join. In fact, replying to some spam messages will simply notify the spamming company that there is a live person on the other end of the e-mail account and encourage more spam. Privacy advocates prefer—and, in fact, many ethical spammers now require—double opt-in, which requires customers to opt-in again to the confirmation message before the spammer can begin sending them e-mail.

Many individuals, privacy advocates, and groups like the Mail Abuse Prevention System (MAPS) have said there should be rules and regulations on how spam is sent, including privacy protections for the recipients. MAPS is a not-for-profit California organization whose mission is to defend the Internet's e-mail system from abuse by spammers, which it does by educating and encouraging ISPs to enforce strong terms and conditions that prohibit ISP customers from engaging in abusive e-mail practices. However, there are still no federal laws regulating unsolicited e-mail and fewer than two-dozen states have addressed this issue. The SueSpammers Web site offers current spamming information on a state-by-state basis. MAPS maintains a blacklist of alleged spammers that many ISPs and corporate e-mail administrators use as a guide to block spam e-mail. Firewalls can be programmed to recognize spam e-mail addresses and prohibit the mail from entering the corporate network. MAPS itself is facing legal action from companies that are trying to get themselves removed from the blacklist.[32]

Carnivore

Today criminals and terrorists rely on e-mails to conduct their trade just like everyone else, and the FBI has developed a surveillance system to take advantage of that fact. **Carnivore** is a highly controversial system used by the FBI to monitor selected e-mail messages—an e-mail version of a telephone wiretap. A device is installed on an e-mail server at a business or ISP to enable Carnivore to track and record e-mail headers that contain the e-mail addresses of recipients and senders, but not the actual contents of any e-mail.

Carnivore enables the FBI to intercept specific e-mails or other computer traffic going to or from specific individuals, while excluding all other computer communications. This system is highly controversial for several reasons. Carnivore opponents insist that law enforcement officials should be required to get the same kind of court order to intercept e-mails as is required for full telephone wiretaps. Instead, the FBI uses Carnivore under the less rigorous rules governing technologies that gather phone numbers dialed by suspects and the number of people calling them. Furthermore, corporate information technology managers and ISPs are concerned that installing Carnivore on their systems may damage their company's information technology hardware and software; compromise the privacy of employees, customers, or suppliers; and create security holes that hackers could exploit. Privacy and civil-rights advocates argue that the use of Carnivore violates the rights of the people to protection against unreasonable search and seizure guaranteed in the United States Constitution's Fourth Amendment. There is even skepticism about the ability of Carnivore to deliver useful results because as of early 2001, none of the information gleaned by the device had found its way into a courtroom.[33]

The FBI made a strong effort to address many of the technical issues by requesting that the Illinois Institute of Technology Research Institute, a not-for-profit research and development organization, evaluate Carnivore. This effort, though, spurred even more controversy. The institute's report, while mostly favorable, pointed out a very disturbing shortcoming—Carnivore cannot eliminate the risk of unauthorized acquisition of electronic communication information by FBI personnel. The report went on to say that due to a lack of audit trails, FBI agents could intercept e-mail for people not listed in the court order and no one could ever know. The inability of the FBI to prove that e-mails not covered by a court order are left alone is cause for great concern.[34]

These findings stimulated many in the technical community to demand that the FBI publish the source code for Carnivore and disclose the details of exactly what it does and how it works. The FBI has refused to do so. Further fueling the controversy are FBI documents that show the agency is planning to add new capabilities to Carnivore, enabling it to reconstruct Web pages that a subject has viewed as well as listen in on voice-over-Internet communications. Without a better understanding of Carnivore, business managers, IT professionals, and ordinary citizens are likely to remain wary and suspicious. However, in the aftermath of the September 11, 2001, attack on the United States, many privacy advocates are having second thoughts about their dislike for this system.

Advanced Surveillance Technology

A number of advances in information technology such as thermal imaging devices, surveillance cameras and face-recognition software, and systems that can pinpoint an individual's position provide exciting new data-gathering capabilities. However, they also lead to a diminishing of individual privacy and add to the question of to what extent technology should be used to capture information about individuals' private lives—where they are, who they are, and what they are doing behind closed doors.

Police can use thermal imaging devices from outside a house to detect patterns of heat being generated from inside. Use of this technology led to the conviction of an Oregon man for growing marijuana in his home. Police used a thermal imager to detect the distinctive heat pattern made by the high-intensity lights that are often used for marijuana cultivation. The police then used this information as the basis for obtaining a search warrant to uncover the contraband.[35]

The police department of Tampa, Florida placed 36 security cameras in the popular Ybor City downtown district and connected them to a powerful computer loaded with face-recognition software. Now everyone who visits the district is subject to having their faces digitally scanned and their noses, cheeks, and chins checked against a mug-shot database of murderers, drug dealers, and other criminals. (Tampa officials used a similar system in January 2001 to scan the crowds at the Super Bowl for possible terrorists; they identified 19 suspects.) Law enforcement officials claim that use of the system does not violate any privacy rights and that its use is no different than having additional policemen walk around trying to identify suspects from mug shots. Besides, they argue, people should expect that their privacy is diminished when they visit the crowded public streets of Ybor City, which are filled with restaurants, nightclubs, stores, and thousands of people. Furthermore, signs in the area warn visitors that "Smart CTTV is in use." Privacy advocates object and say use of such systems amounts to putting the public in a digital police lineup.[36]

Following the September 11, 2001, terrorist attack, there has been much interest in the use of cameras posted at airport check-ins to analyze each passenger's face and match its characteristics to a database of terrorists and wanted felons. For such measures to work, though, the database must contain mug shots of the terrorists, meaning someone must already be under suspicion or previously arrested in order to be identified through this system. As September 11 has proven, it isn't always clear who the bad guys are.

The widespread use of cell phones has created another opportunity for advanced surveillance technology. The Federal Communications Commission (FCC) has asked cell-phone companies to implement methods for identifying the location of all users so that the police, fire, and medical professionals can be accurately dispatched to 911 callers. To achieve this end, cell-phone companies are placing Global Positioning Systems (GPS) chips into some new cell phones. Similar location tracking technology is also available for personal digital assistants, laptop computers, autos, trucks, and boats.

As a consequence, banks, retailers, and airlines are anxious to gain real-time access to consumer location data and have already devised a number of new services they wish to provide—sending digital coupons for stores to which consumers are near, providing the location of the nearest ATM, and updating travelers on flight and hotel information. The airlines are considering the use of wireless devices to enable passengers to check in for flights when they are within a certain distance of the gate and then to monitor when the person passes through the gate.

Businesses claim that they will respect the privacy of wireless users and allow them to "opt-in" or "opt-out" of marketing programs that are based on their location data. Wireless spamming is a distinct possibility—a user might continuously receive wireless ads, notices for local restaurants, and shopping advice while walking down the street. Another concern is that the data could be used to track people down at any time or to figure out where they had been at some particular instant. The potential for loss of privacy of one's location when using a cell phone will cause some people to reconsider when and how they use a cell phone in the future.

Summary

1. **What is the right of privacy and what is the basis for the protection of an individual's privacy under the law?**

   As a legal concept, the right to privacy has four aspects: protection from unreasonable intrusion upon one's isolation, protection from appropriation of one's name or likeness, protection from unreasonable publicity given to one's private life, and protection from publicity that unreasonably places one in a false light before the public.[37]

   A number of legislative actions have been taken over the past 40 years that affect an individual's privacy; most of these actions address invasion of privacy by the government rather than by private industry. In addition, there is no overarching national data privacy policy, so current legislation is sometimes inconsistent and even conflicting. This legislation includes The Communications Act of 1934, The Freedom of Information Act of 1966, The Privacy Act of 1974, The Electronic Privacy Act of 1986, the 1994 Communications Assistance for Law Enforcement Act, The Telecommunications Deregulation and Reform Act of 1996, the Children's Online Privacy Protection Act of 1998, The European Community Directive 95/46/EC of 1998, the 1998 Gramm-Leach-Biley Act, and the 2001 Patriot Act.

2. **What are the two fundamental forms of data encryption and how does each work?**

   Data encryption is an essential tool for ensuring confidentiality, integrity, and authenticity of messages and business transactions. A public key encryption system uses two keys to encode and decode messages. One key of the pair, the message receiver's public key, is readily available to the public and is used by anyone to send that individual encrypted messages. The second key, the message receiver's private key, is kept secret and is known only by the message receiver. Its owner uses the private key to decrypt messages. A private key encryption system uses a single key to both encode and decode messages. Both the sender and receiver must know the key to communicate. It is critical that no one else learns the key or else all messages between the two can be decoded by others.

3. **What are the various strategies for consumer profiling and what are the associated privacy issues?**

   Marketing firms capture data from numerous sources to build databases detailing a large amount of consumer behavior data. Online marketers can place an electronic cookie on a consumer's hard drive so that they can recognize the consumer when he or she returns to a Web site. It can also be used to store information about the person. Marketers employ personalization software to analyze browsing habits to deduce personal interests and preferences. The use of cookies and personalization software is controversial because it enables companies to find out information about consumers and to potentially share it with other companies, without consumers' explicit permission.

4. **What must an organization do to treat consumer data responsibly?**

   A widely accepted approach to treating consumer data responsibly is the adoption of the Code of Fair Information Practices and the 1980 OECD privacy guidelines. Under these guidelines, an organization collects only the personal information that is necessary to deliver its product or service. The company ensures that the information is carefully protected and that it is accessible only by those with a need to know, and it provides a process for consumers to review their own data and make corrections. In addition to adopting the OECD guidelines, many companies are appointing CPOs or other senior managers to oversee their data privacy policies and initiatives.

5. **Why and how are employers increasingly implementing workplace monitoring?**

   Employers are increasingly recording and reviewing employee communications and activities on the job, including monitoring workers' phone calls, e-mail, Internet connections, and computer files. Our society is struggling to define the extent to which employers should be able to monitor employees. On the one hand, employers must be able to guarantee a work environment conducive to all workers, ensure a high level of worker productivity, and avoid the costs of defending against "nuisance" lawsuits. On the other hand, privacy advocates want federal legislation that keeps employers from infringing upon the privacy rights of employees.

6. What are the capabilities of the Carnivore system and other advanced surveillance technologies, and what privacy issues are raised by these technologies?

Carnivore is a highly controversial system used by the FBI to monitor selected e-mail messages—an e-mail version of a telephone wiretap. Carnivore opponents insist that law enforcement officials should be required to get the same kind of court order to intercept e-mails as is required for full telephone wiretaps.

Instead, the FBI uses Carnivore under the less-rigorous rules governing technologies that gather phone numbers dialed by suspects and the numbers of people calling them. In addition, privacy and civil-rights advocates argue that the use of Carnivore violates the rights of the people to protection against unreasonable search and seizure guaranteed in the United States Constitution's Fourth Amendment.

A number of advances in information technology such as such as thermal imaging devices, surveillance cameras and face-recognition software, and systems (like GPS) that can pinpoint an individual's position provide exciting new data-gathering capabilities. However, they also lead to a diminishing of individual privacy and add to the question of to what extent technology should be used to capture information about individuals' private lives—where they are, who they are, and what they are doing behind closed doors.

Review Questions

1. What role do the BBB Online and TRUSTe play in protecting consumer privacy?
2. What is opt-in? What is opt-out?
3. What is consumer profiling?
4. Why do marketers engage in consumer profiling?
5. What benefits can consumer profiling provide to the consumer?
6. What is a cookie and how is it used?
7. What three kinds of data about individuals are gathered as they surf the Internet?
8. What actions can be taken to prevent the deposit of cookies on a hard drive?
9. Why is it said that public workers enjoy greater privacy protection than private industry workers?
10. Why is spam sometimes considered objectionable? What are some legitimate business benefits of using spam to reach potential customers?
11. What is the FOIA? Why do whistle blowers frequently use it?
12. What are the privacy guidelines set by the OECD? Why are they considered significant?
13. What is data encryption? What is the AES?
14. What is the role of a CPO?
15. Briefly distinguish between private key and public key encryption.

Discussion Questions

1. Compare and contrast the European Union and United States philosophy on data privacy. How might the EU system be adapted to American consumer/business needs and expectations?
2. What benefits can consumer profiling provide to you as a consumer? Do these benefits outweigh the loss of your privacy?
3. What are three types of personalization software? Which do you think is most effective? Why?
4. An FOIA exemption prevents disclosure of records if that disclosure would cause an invasion of someone's personal privacy. Develop a hypothetical example where clearly the individual's privacy interests at stake are outweighed by competing public interests. Develop a second hypothetical example where clearly the individual's privacy interests at stake are not outweighed by competing public interests.
5. State the Fourth Amendment of the U.S. Constitution. Does it apply to the actions of private industry?
6. Are you supportive of the CALEA? Why or why not?
7. Much of the world is "at war" against terrorism. Have you seen any examples that this has caused a decrease in our right to privacy? Is such action justifiable?
8. Why do employers monitor workers? Do you think they should be able to do so? Why or why not?
9. Do you think that law enforcement agencies should be able to use the Carnivore system? Why or why not?

What Would You Do?

1. As the information systems manager for a small manufacturing plant, you are responsible for all aspects of the use of information technology. A new inventory control system is being implemented to track the quantity and movement of all finished products stored in a local warehouse. Each time a fork-lift operator moves a case of product, he or she must first scan the UPC code on the case. Not only is the product information captured, but also the day, time, and fork-lift operator identification. This data is transmitted over a LAN to the inventory control computer that then displays information about the case and where it should be placed in the warehouse.

   The warehouse manager is excited about using the case movement data to monitor the productivity of the workers. He will be able to tell how many cases per shift each operator moves, and he plans to use this data to provide performance feedback that could result in pay increases or termination. He has asked you if there are any potential issues with using the data in this manner, and, if so, what should be done to avoid them. How would you respond?

2. As a young and highly successful member of your company's marketing organization, you have been asked to take on the role of your company's first ever CPO. What questions would you ask before accepting this role? Assume that you have agreed to become the CPO. One of your first actions is to develop a comprehensive data privacy policy. How would you go about doing this?

3. You are a new marketing manager for the Ford Motor Company. You are considering the use of spam to promote the latest and greatest automobile model that is targeted to young, affluent adults. List the advantages and disadvantages of such a marketing strategy. Would you recommend this means of promotion? Why or why not?

4. You are the CPO of a medium-sized manufacturing company with sales of over $250 million per year with almost $50 million coming from Internet-based sales. You have been challenged by the vice president of sales to change the company's Web site data privacy policy from opt-in to opt-out and to allow the sale of customer data to other companies. The vice president has estimated that this change would bring in at least $5 million per year in added revenue with little additional expense. How would you respond to this request?

Cases

1. HIPAA and the Upcoming Furor over Medical Records

The goal of the Health Insurance Portability and Accountability Act of 1996 (HIPAA) is to require health care organizations to implement cost-effective procedures for exchanging medical data. Health care organizations must employ standard electronic transactions, codes, and identifiers designed to enable them to fully "digitize" medical records and make it possible to use the Internet rather than expensive private networks for electronic data interchange. The Department of Health and Human Services developed over 1500 pages of specific rules (see *www.hhs.gov*) governing the exchange of such data, with a compliance deadline of April 2003. The regulations affect 1.5 million health care providers, 7,000 hospitals, and 2,000 health care plans.

Under the HIPAA provisions, health care providers must obtain written consent from patients for the use or disclosure of information in their medical records. Patients are also guaranteed the right to inspect and copy their medical records and to suggest changes to correct inaccuracies. Health care providers must keep track of everyone who received medical information from a patient's medical file. Patients can demand that doctors and hospitals provide an accounting of all disclosures spanning the past six years. These provisions will affect every doctor, patient, hospital, pharmacy, and insurer.

Health care companies must appoint a privacy officer to develop privacy policies and procedures as well as to train employees in how to handle sensitive data. These actions must address the potential for unauthorized access to data by outside hackers as well as the more likely threat of internal misuse of data. It is the employee within the health care organization who is much more likely to compromise confidentiality, either intentionally or accidentally. For example, during an upgrade to one company's information systems, hundreds of e-mail messages containing sensitive information were accidentally sent to members of a large HMO. Not only was there a potential loss of privacy due to the messages being intercepted by people who were not the intended recipients, but also each recipient lost some personal privacy by simply having his or her name appear on the distribution list for the message. Health care companies must also guard against the prospect of personnel with authorized access using data inappropriately—such as a cardiologist reading a patient's psychiatric records online and telling her that her chest pains are not real but related to her psychosis.

HIPAA assigns responsibility to health care organizations, as the originators of individual medical data, for certifying that their business partners (billing agents, insurers, debt collectors, research firms, government agencies, and charitable organizations) also comply with HIPAA security and privacy rules. This provision of HIPAA has health care executives especially concerned as they do not have direct control over the systems and procedures that their partners implement. Those who misuse data may be fined $250,000 and serve up to 10 years in prison.

As the full details of HIPAA have become better understood, many experts have become concerned. Some fear that between the increasing demands for disclosure of patient information and the impending full digitization of medical records, patient confidentiality will be lost. Many think that the HIPAA provisions are too complicated and will miss the original objective of reducing medical industry costs and instead increase costs and paperwork for doctors without improving medical care. All agree that the medical industry must make a substantial investment to achieve compliance. Government experts estimate that it will cost each hospital between $100,000 and $250,000 to comply with HIPAA's data privacy and security regulations. Meanwhile, a study by Blue Cross/Blue Shield puts the costs much higher—$775,000 to $6 million per hospital.

The Agency for Healthcare Research and Quality (the research arm of The Department of Health and Human Services) states that HIPAA will require computer systems that can greatly reduce, if implemented correctly, the adverse reactions caused by medication errors. The agency estimates that hospitals will save $500,000 in direct costs annually.

Questions:
1. What are the potential benefits from full implementation of HIPAA—from a patient's perspective and from a health care organization's perspective?
2. What actions could a privacy officer take to be able to certify that a health care organization's business partners also comply with HIPAA security and privacy rules?
3. What do you see as the likely negative effects of HIPAA? How well do you think these effects balance against the benefits?

Sources: adapted from Tracy Mayor, "The Privacy Problem," *CIO*, January 15, 2001, pp. 75–84; Robert Pear, "Medical Industry Lobbies to Rein in New Patients Privacy Rules," *The New York Times on the Web*, February 12, 2001, accessed at www.nytimes.com; Robert Pear, "Health Secretary Delays Medical Records Protections," *The New York Times on the Web*, February 12, 2001, accessed at www.nytimes.com; George V. Hulme, Diane Rezendes, and Khir Allah, "Protecting Privacy," *Information Week*, April 16, 2001, pp. 22–24; Clinton Wilder and John Soat, "The Ethics of Data," *Information Week*, May 14, 2001, pp. 37–48.

2. Echelon—Top Secret Intelligence System

Echelon is a top-secret electronic eavesdropping system managed by the National Security Agency (NSA) of the United States and known to be used by the intelligence agencies of England, Canada, Australia, and New Zealand. It is capable of intercepting and decrypting almost any electronic message sent anywhere in the world via satellite, microwave, cellular, or fiber-optic telecommunications, including radio and TV broadcasts, phone calls, computer-to-computer data transmission, faxes, and e-mail. It may have been in operation since as early as the 1970s, but it wasn't until the 1990s that journalists using the FOIA were able to confirm its existence and gain insight into its capabilities. Although Echelon is the world's largest and most sophisticated surveillance network, it is by no means the only one. Russia, China, Denmark, France, the Netherlands, Russia, and Switzerland operate Echelon-like systems to obtain and process intelligence by listening in on electronic communications.

Which electronic transmissions are captured and what Echelon is able to do with the messages is subject to much conjecture. Even if all electronic messages worldwide were unencrypted, finding those messages that warranted further attention would be an enormous, computer-intensive task. As a result, it is likely that Echelon targets communications to and from specific individuals and organizations rather than trying to assimilate all electronic messages. Thus, some subset of all possible messages is forwarded to the massive United States intelligence operations at Fort Meade, Virginia, where powerful computers look for code words or key phrases among the messages. Intelligence analysts peruse any conversation or document thus flagged by the system, and significant messages are then forwarded to the agency that requested the information.

A number of intelligence satellites in orbit are used to detect signals that normally dissipate into space—radio signals, mobile phone conversations, and microwave transmissions. In addition, at least six ground-based stations throughout the world are used to monitor the communication satellites of Intelsat, the world's largest commercial satellite communications services provider.

Computer processing speeds and the science of speech recognition probably are not yet advanced enough for a real-time global listening system capable of transcribing the hundreds of thousands of calls that are happening at any instant in time. However, Echelon is capable of voice pattern matching and can identify who is speaking if their voice pattern is stored in its database. Also, it employs recording systems that are capable of automatically triggering tape recordings based on "hearing" key words.

Echelon employs special software and speech recognition technology to convert any audio communications into formatted, searchable text. A half-hour broadcast can be processed and stored in searchable format in 10 minutes. Currently the software understands only American English, but the CIA is enhancing it to handle Chinese and Arabic. Other Echelon software is used to alert intelligence analysts any time a new page goes up on a Web site of interest. CIA personnel use special software to perform searches in English of Web sites developed in Chinese, Japanese, Russian, and eight other languages. The software then translates the text of the Web site into English.

This immense, highly sophisticated surveillance system apparently operates with little oversight, and the various agencies that run Echelon have provided few details as to the legal guidelines governing the project. Indeed, the governments of the countries believed to be involved have failed to officially acknowledge the existence of Echelon. Because of this, there is no way of knowing its true capabilities and exactly how it is being used.

Echelon intercepts both sensitive government data and corporate information. It also provides the opportunity to illegally spy on private citizens. It is no wonder that privacy advocates are upset with the secrecy surrounding the system and its great potential for misuse. They feel that Echelon can be directed against virtually any citizen in the world with the full knowledge and cooperation of their government.

In the U.K, Echelon has already been accused of spying on organizations such as Amnesty International—an international organization that seeks to ensure fair and prompt trials for political prisoners and that opposes human rights abuses. In addition, in September 1999, the European Union released a report highly critical of the operators of Echelon for using it to intercept confidential company information and divulging it to favored competitors to help win contracts. The report alleged that Airbus Industrie of France lost valuable contracts because information intercepted by Echelon was forwarded to the Boeing Company to help it obtain a competitive advantage.

In the United States, the ACLU and others are concerned that Echelon may be used without a court order to intercept communications involving Americans. The Foreign Intelligence Surveillance Act prohibits interception of certain communications for intelligence purposes without a court order unless the Attorney General certifies that certain conditions are met. These conditions include a limitation that "there is no substantial likelihood that the surveillance will acquire the contents of any communication to which a United States person is a party."

Echelon supporters know that communications surveillance is successful in gathering enemy intelligence and was a key to the success of the allied military effort in World War II. They also argue that tragedies such as the September 11, 2001, attack and the bombing of the federal building in Oklahoma City are proof that such a surveillance system is necessary to forewarn authorities and potentially prevent major terrorist activities. In that regard, the United States agreed to share highly classified material from Echelon with the Spanish government to aid in its battle against the Basque separatist group ETA. As a result, the Spanish are now receiving decoded intercepts relating to the ETA's plans for terrorist operations.

Questions:

1. Are you for or against the use of the Echelon for eavesdropping on electronic communications? Why or why not? Is your opinion affected by the September 11, 2001, terrorist attacks?

2. Develop a set of plausible conditions under which the directors of Echelon would authorize the use of the system to listen to specific electronic communications.

3. What sort of expanded or new capabilities might Echelon have 10 years from now as information technology continues to improve at a rapid pace? What additional privacy issues might be raised by these new capabilities?

Sources: adapted from "Echelon Surveillance System Threatens U.S. Privacy," *Online Newsletter*, September 2000, accessed at www.findarticles.com; Mathew Schwartz, "Intercepting Messages," *Computerworld*, August 28, 2000, pp. 48–49; Isambard Wilkinson, "US Wins Spain's Favour With Offer to Share Spy Network Material," *The Guardian; The Telegraph*, June 18, 2001, accessed at www.smh.com.au; John Lettice, "French Echelon Report Says Europe Should Lock Out US Snoops," *The Register*, October 13, 2000, accessed at www.theregister.co.uk; Veernon Loeb, "Making Sense of the Deluge of Data," *The Washington Post*, March 26, 2001, Page A23.

Endnotes

[1] Robert Scheer, Privacy In the Digital Age," *Yahoo! Internet Life*, October 2000, pg. 101.

[2] *The Privacy Journal* at www.townonline.com/specials/privacy/, accessed on July 19, 2001.

[3] John R. Boatright, *Ethics and the Conduct of Business*, Third Edition, © 2000, Prentice Hall: Upper Saddle River, NJ, pp. 166–168.

[4] "About" from the OECD Web site at www.oecd.org, accessed on June 5, 2002.

[5] Rebecca Lynch, "Privacy," *CIO*, October 1, 2000, pg. 208.

[6] "What's Next for COPPA," *PR Newswire*, April 18, 2001.

[7] Linda Rosencrance, "FTC Fines Kids Sites for Privacy Violations," *Computerworld*, April 23, 2001, accessed at www.computerworld.com.

[8] Scarlett Pruitt, "Supreme Court Hears Arguments on Net Porn Law," *Computerworld*, November 28, 2001, accessed at www.computerworld.com.

[9] Rebecca Lynch, "Privacy," *CIO*, October 1, 2000, pg. 208.

[10] Ibid.

[11] Brock N. Meeks, "Is Privacy Possible in the Digital Age," *MSNBC News*, December 7, 2000.

[12] Ibid.

[13] Mike France, "Are Lenders Playing Fair with Your Privacy?", *Business Week Online*, April 2, 2001, accessed at www.businessweek.com.

[14] Eric Schmitt, "Ashcroft Proposes Rules for Foreign Visitors," *The New York Times on the Web*, June 5, 2002, accessed at www.nytimes.com.

[15] Peter Wayner, "Attacks on Encryption Code Raise Questions About Computer Vulnerability," *The New York Times on the Web*, January 5, 2000, accessed at www.nytimes.com.

[16] John Schwartz, "U.S. Selects a New Encryption Technique," *The New York Times on the Web*, October 3, 2000, accessed at www.nytimes.com.

[17] John Schwartz, "U.S. Selects a New Encryption Technique," *The New York Times on the Web*, October 3, 2000, accessed at www.nytimes.com.

[18] Paulina Borsook, "The Uses and Abuses of Customer Profiling," *Knowledge Management*, November 1999, pp. 57–63.

[19] Robert Scheer, "Privacy In the Digital Age," *Yahoo! Internet Life*, October 2000, pg. 101.

[20] Matt Hicks, "Getting Personal," *eWeek*, October 2, 2000, pp. 61–64, 71.

[21] Patrick Thibodeau, "Online Profiling," *Computerworld*, September 18, 2000, pg. 56.

[22] Daintry Duffy, "You Know What They Did Last Night," *CIO*, April 1, 2000, pp. 188–196.

[23] Will Rodger, "DoubleClick Backs Off Web-Tracking Plan," *USA Today*, March 2, 2000, accessed at www.usatoday.com.

24 Jennifer Disabatino, "Disney: We'll Retire Toysmart Customer List," *Computerworld*, July 17, 2000, pg. 10.

25 Simson L. Garfinkel, "Opinion—Difference Engine, The Social Impact of Technology," *CIO*, June 1, 2000, pp. 178–180.

26 Kim S. Nash, "Chief Privacy Officers: Forces or Figureheads?", *Computerworld*, November 13, 2000, pp. 62–65.

27 Kim S. Nash, "Chief Privacy Officers: Forces or Figureheads?", *Computerworld*, November 13, 2000, pp. 62–65.

28 Betsy Stark, "Every Step you Take," *ABC News*, January 4, 2001, accessed at http://search.abcnews.go.com.

29 Kenneth A. Kovach, "Electronic Communication in the Workplace—Something's Got to Give," *Business Horizons*, July 2000, accessed at www.findarticles.com.

30 Jennifer Disabatino, "Marketing's Beef with Spam Labels," *Computerworld*, November 21, 2000, accessed at www.computerworld.com.

31 Todd R. Weiss, "AOL Sues Company Over Alleged E-mail Spamming," *Computerworld*, January 4, 2001, accessed at www.computerworld.com.

32 Todd R. Weiss, "AOL Sues Company Over Alleged E-mail Spamming," *Computerworld*, January 4, 2001, accessed at www.computerworld.com.

33 Larry Kahaner, "Hungry for Your E-mail," *Information Week*, April 23, 2001, pp. 59–64.

34 Larry Kahaner, "Hungry for Your E-mail," *Information Week*, April 23, 2001, pp. 59–64.

35 Linda Greenhouse, "Justices Look at Heat-Seeker's Ability to Pierce the Home," *The New York Times on the Web*, February 22, 2001, accessed at www.nytimes.com.

36 Dana Canedy, "Tampa Scans the Faces in Its Crowds for Criminals," *The New York Times on the Web*, July 4, 2001, accessed at www.nytimes.com.

37 John R. Boatright, *Ethics and the Conduct of Business*, Third Edition, © 2000, Prentice Hall: Upper Saddle River, NJ, pp. 166–168.

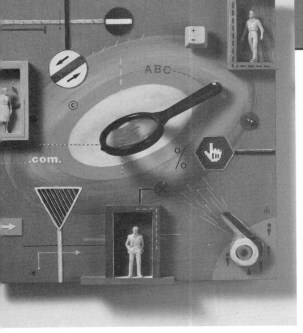

Chapter 5

FREEDOM
of Expression

"Society in every state is a blessing, but Government, even in its best state,
is a necessary evil; in its worst state, an intolerable one."
— Thomas Paine, American Revolutionary patriot in "Common Sense,"
published 1776.

After an Intel engineer was fired in 1996 following a dispute over work-related injuries, he started a Web site to publicize what he claimed was Intel's mistreatment of workers. He also sent e-mail messages to at least 8,000 Intel employees on six separate occasions over a two-year period starting in December 1996 to publicize his Web site and ask recipients to send in complaints about the company.

Intel sought legal action to put an end to the e-mail. In April 1999, a Sacramento superior court judge ruled that the former employee's actions constituted trespassing onto Intel's private computer system and that "the mere connection of Intel's e-mail system to the Internet does not convert it into a public forum." The judge further ruled that the e-mails were not protected speech and Intel was granted a permanent injunction to keep out e-mail sent by the former employee.

This ruling could set a precedent that allows companies and Internet Service Providers (ISPs) to block out anyone they want for any reason, according to a spokesperson for the Electronic Frontier Foundation (EFF), an online free speech group. A spokesperson for Intel pointed out that the company had done nothing to stop publication of the former employee's complaints on his Web site, nor did it try to stop him from protesting on college campuses and at Intel plant and office sites.

The employee appealed the case with support from the EFF and the American Civil Liberties Union (ACLU). In December 2001, a California appeals court also upheld the superior court's decision and said that the employee's sending of e-mail was equivalent to trespassing. The employee and his attorneys then appealed the case to the California Supreme Court, which agreed to hear the case in May 2002. As of this writing, the case has not yet been decided.

Sources: adapted from Malcolm Maciahlan, "Intel Wins E-mail Case," Tech Web, April 29, 1999, accessed at www.techweb.com; Cheri Paquet, "Intel Lawsuit Calls Barring E-mail Fair Game," CNN.com, May 3, 1999, accessed at www.cnn.com; "Intel's Attempt to Muzzle Ex-Worker's E-mails Violates First Amendment, ACLU Says," May 12, 2000, ACLU Web site, at www.aclu.org; Declam McCullagh, "Axed Intel Man Loses E-mail Case," Wired News, December 12, 2001, accessed at www.wired.com; Scarlet Pruitt, "California High Court to Hear Mass E-mail Case," Computerworld, March 28, 2002, accessed at www.computerworld.com.

As you read this chapter, consider the following questions:

1. What is the legal basis for the protection of our freedom of speech and what types of speech are not protected under the law?

2. In what ways does the Internet present new challenges in the area of freedom of expression?

3. What are the key freedom of speech issues that must be addressed in considering the use of information technology?

FIRST AMENDMENT RIGHTS

The Internet enables a worldwide exchange of news, ideas, opinions, rumors, and information. Indeed, the very strengths of the Internet—its broad accessibility, open-minded discussions, and anonymity make it an ideal communications medium. One of the most powerful aspects of the Internet is that it provides an easy and inexpensive way for a speaker to send a message indiscriminately to a large audience, potentially thousands of people worldwide. In addition, given the right e-mail addresses, a speaker can aim a message with laser accuracy at a select subset of powerful and influential people.

Individuals must often make ethical decisions in regard to how they will use such remarkable freedom and power. Organizations and governments have attempted to establish policies and laws to help guide individuals as well as to protect their own interests. Business organizations, in particular, have sought to conserve corporate network capacity, avoid legal liability, and improve worker productivity by limiting the non-business use of information technology resources.

The right to freedom of expression is one of the most important rights for free people everywhere. The First Amendment to the United States Constitution was adopted to guarantee this and specific other rights. Over the years, a number of federal, state, and local laws have been found to be unconstitutional because they violated one of the tenets of this amendment.

The First Amendment of the United States Constitution reads as follows: "Congress shall make no law respecting an establishment of religion, or prohibiting the free exercise thereof; or abridging the freedom of speech, or of the press; or the right of the people peaceably to assemble, and to petition the government for a redress of grievances." In other words, the First Amendment protects the right to freedom of religion and freedom of expression from government interference. This amendment has been interpreted by the Supreme Court to apply to the entire federal government—even though it only expressly applies to Congress.

Numerous court decisions have broadened the definition of speech to include non-verbal, visual, and symbolic forms of expression, such as the burning of the flag, dance movements, and hand gestures. Sometimes the "speech" at issue is unpopular or highly offensive to the majority of people; however, the Bill of Rights provides protection for the views of the minority. The Supreme Court has also ruled that the First Amendment protects the right to speak anonymously as part of the guarantee of free speech.

The Supreme Court has held that the following types of speech are not protected by the First Amendment and may be totally forbidden by the government: obscene speech, defamation, incitement of panic, incitement to crime, "fighting words," and sedition (incitement of discontent or rebellion against a government). Two of these types of speech, obscene speech and defamation, are particularly relevant to information technology.

Obscene Speech

Miller v. California is the Supreme Court case that established a test to determine if material is obscene and therefore not protected by the First Amendment. Miller, after conducting a mass mailing campaign to advertise the sale of adult material, was convicted of violating a California statute prohibiting the distribution of obscene material. Some unwilling recipients of Miller's brochures complained to the police, initiating the legal proceedings. Miller's conviction was specifically based on his conduct in causing five unsolicited advertising brochures to be sent through the mail. Although the brochures contained some descriptive printed material, they primarily consisted of pictures and drawings explicitly depicting men and women engaging in sexual activity. In its 1973 ruling for the defendant in *Miller v. California* the Supreme Court (in a narrow 5-4 ruling) determined that speech is obscene and is not protected under the First Amendment when it meets the three following conditions: 1) Would the average person, applying contemporary community standards, find that the work, taken as a whole, appeals to the prurient interest? 2) Does the work depict or describe, in a patently offensive way, sexual conduct specifically defined by the applicable state law? and 3) Does the work, taken as a whole, lack serious literary, artistic, political, or scientific value? These three tests have become the United States standard for determining if something is obscene. The requirement that the work be assessed by its impact on the average adult member of the community has raised many issues—who is an average person, what are the contemporary community standards and, in the case of potentially obscene material displayed worldwide on the Internet, what is the community?

Defamation

Your right to freedom of expression is restricted when your expressions, whether spoken or written, are untrue and cause harm to another person. The publication of a statement of alleged fact that is false and that harms another person is **defamation**. The harm is often of a financial nature, in that it reduces the individual's ability to earn a living, work in a profession, or run for an elected office. An oral defamatory statement is **slander**. A written defamatory statement is **libel**. Because defamation is defined as an untrue statement of fact, truth is an absolute defense to a charge of defamation. The expression of one's opinion is not considered defamation and usually no action can be taken against someone for expressing their opinion. However, although individuals have the right to express their opinions, they must exercise care in their Internet communications to avoid possible charges of defamation. Organizations must also be on guard and prepared to take action in the event of libelous attacks against them.

KEY FREEDOM OF EXPRESSION ISSUES

Information technology has provided amazing new ways to communicate with people around the world. With these new ways come new responsibilities and new ethical issues. This section will discuss a number of key issues related to freedom of expression, including controlling access to information on the Internet, anonymity, defamation and hate speech, and pornography.

Controlling Access to Information on the Internet

Although there are clear and convincing arguments to support freedom of speech on the Internet, the issue becomes problematic when one considers the ease with which children can access the Internet. Indeed, many free speech advocates acknowledge the need to somehow restrict children's access. The conundrum is that it is difficult to restrict Internet access by children without at the same time restricting access by adults. In an attempt to address this issue, the United States government has passed laws, and software manufacturers have invented special software, with the goal of blocking access to objectionable material.

The Communications Decency Act (CDA)

The Telecommunications Deregulation and Reform Act became law in 1996. Its purpose was to allow freer competition among phone, cable, and TV companies.

Embedded in the Telecommunications Act was the Communications Decency Act (CDA), aimed at protecting children from online pornography. The CDA imposed $250,000 fines and prison terms of up to two years for the transmission of "indecent" material over the Internet.

The *Reno v. ACLU* suit, filed in February 1996, challenged the criminalizing of so-called "indecency" on the Internet. The government appealed the case to the Supreme Court after a federal, three-judge panel ruled unanimously that the law unconstitutionally restricts free speech. Examples of indecency identified as potentially criminal by government witnesses included Internet postings of the photo of the actress Demi Moore naked and pregnant on the cover of *Vanity Fair* and any use online of the infamous "seven dirty words." The plaintiffs included the ACLU, Planned Parenthood, Stop Prisoner Rape, Human Rights Watch, and Critical Path AIDS Project. Many of these organizations feared that much of their online material (examples cited included speech about abortion, prisoner rape, safe sex practices, and many other sexually related topics) could be classified as indecent. They argued that such information was important to both minors and adults.[1]

The problem with the CDA was its broad language and vague definition of "indecent," a standard left to individual communities to decide. In June 1997, the Supreme Court ruled the law unconstitutional and declared that the Internet must be afforded "the highest protection available under the First Amendment."[2] The Supreme Court said in its ruling that "the interest in encouraging freedom of expression in a democratic society outweighs any theoretical but unproven benefit of censorship." The Supreme Court went on to say that "the growth of the Internet has been and continues to be phenomenal. As a matter of constitutional tradition, and in the absence of evidence to the contrary, we presume government regulation of the content of speech is more likely to interfere with the free exchange of ideas than to encourage it."[3]

If the CDA had been judged constitutional, it would have opened up all aspects of Internet content to legal scrutiny. It is likely that many current Web sites would either be non-existent or look much different today had the law not been overturned. In addition, webmasters and Web site owners whose sites might be deemed indecent under the CDA would be operating under an extreme risk of personal liability.

Internet Filtering

An **Internet filter** is software that can be installed on a personal computer along with a Web browser to block access to certain Web sites that contain inappropriate or offensive material. Network administrators may choose to install filters on employees' computers in order to prevent them from viewing sites that contain pornography or other objectionable material. Employees unwillingly exposed to such material would have a strong case for sexual harassment. The use of filters can also ensure that employees do not waste their time viewing non-business oriented Web sties. There are many Internet filters available, including Net Nanny, Cybersitter, Cyber Patrol, SurfGuard, and SurfWatch. HateFilter™ is an Internet filter that can be downloaded from Anti-Defamation League's (ADL) Web site. This filter contains a redirect feature that offers users who try to access a blocked site advocating bigotry, hatred, or violence towards groups on the basis of their ethnicity, race, religion, or sexual orientation the chance to link directly to related ADL educational material.

The Internet Content Rating Association (ICRA) is a nonprofit organization whose members include Internet industry leaders such as AOL/Time Warner, Bell South, British Telecom, IBM, Microsoft, UUNet, and Verizon. ICRA's mission is to enable the public to make informed decisions about electronic media through the open and objective labeling of content. Its specific goals are to protect children from potentially harmful material while at the same time safeguarding free speech on the Internet.

With the ICRA rating system, Web authors fill out an online questionnaire describing the content of their site in terms of what is and isn't present. The broad topics covered are as follows: chat, the language used on the site, the nudity and sexual content of a site, the violence depicted on the site, and other content areas such as alcohol, drugs, gambling, and suicide. Within each broad category, the web author is asked questions about whether a specific item or feature is present or absent on the site. Based on the author's responses, ICRA then generates a content label (a short piece of computer code) that the author adds to the site. These content labels generated by ICRA conform to an Internet industry standard known as the Platform for Internet Content Selection (PICS). Internet users can then set their browser to allow or disallow access to Web sites based on the objective information declared in the content label and their own subjective preferences.

Note that the ICRA does not rate Internet content—the content providers do. However, reliance on Web site authors to do their own rating has its weaknesses. For one, many hate and sexually explicit sites don't have an ICRA rating. They won't be blocked unless a browser is set to block all unrated sites, which leads to blocking so many actually acceptable sites that it can make Web surfing a useless activity. Site labeling also depends on the honesty with which Web site authors rate themselves. Web site authors can lie when completing the ICRA questionnaire so that their site receives a content label that doesn't accurately reflect the site's content. For these reasons, at this time, site labeling is at best a complement to other filtering techniques.

Another approach to restricting access to Web sites is to subscribe to an ISP that performs the blocking itself. The blocking occurs at the ISP's server, as opposed to using software loaded onto each user's computer. ClearSail is such an ISP and prevents access to known Web sites that address the following topics: bomb making, gambling, hacking, hate, illegal drugs, pornography, profanity, Satan, and suicide. ClearSail employees search the Internet each day to uncover new Web sites to add to ClearSail's block list. It blocks known URLs, known pornographic hosting services, keywords in URLs, and search keywords. There are millions of blocked Web pages as a result of ClearSail's filtering capability. Additionally, newsgroups are blocked because of the uncontrollable amount of pornography contained within them.

Children's Internet Protection Act

In December 2000, Congress passed the Children's Internet Protection Act, which required federally financed schools and libraries to use some form of technology protection measure (such as an Internet filter) to block access to obscene material, pornography, and anything considered harmful to minors. Congress did not specifically define which content or Web sites should be forbidden and which measures should be used—this was left to individual school districts and library systems to

117

decide. Any school or library that failed to comply with the law would no longer be eligible to receive federal money from a number of sources.

Opponents of the law fear that it transferred power over the education process to private software companies who develop the Internet filters and define which sites are to be blocked. Furthermore, the motives of these companies are unclear—some of the filtering companies track students' Web surfing activities and sell the data to market research firms. Opponents also point out that current versions of these filters are ineffective. They block access to legitimate sites while at the same time allowing users to access objectionable sites. Yet another objection is that the penalties associated with the act could cause schools and libraries to lose federal funds from the e-Pay program, which is designed to help pay for Internet connections. Loss of federal funds would lead to the creation of a reduced-capability version of the Internet for students at poorer schools, which have the fewest alternatives to federal aid. The end result would be that the law would counteract the e-Pay program, which was intended to help bridge the so-called digital divide in America between the rich and poor and the urban and rural.

Proponents of the Children's Internet Protection Act contend that shielding children from drugs, hate, Satan, pornography, and other topics is a sufficiently compelling reason to justify the imposition of filters. They argue that current Internet filters are highly flexible and customizable and that their critics exaggerate their limitations. They point out that schools and libraries can still elect not to implement a children's Internet protection program; they just won't receive federal money for Internet access.

Many school districts have implemented programs consistent with the Children's Internet Protection Act. Acceptance of an Internet filtering system is improved if the system and its rationale are first discussed with parents, students, teachers, and administrators. Then the program is refined, taking into account everyone's feedback. An essential element of a successful program is to require that students, parents, and adult employees sign an agreement outlining the school district's acceptable use policies for accessing the Internet. Controlling Internet access from one central school district network, rather than letting each individual school set up its own filtering system, reduces administrative effort and ensures consistency. Importantly, procedures must be defined to block new, objectionable sites as well as to remove blocks from Web sites that should be accessible.

Implementing the Children's Internet Protection Act in libraries is much more difficult because their services are open to people of all ages, including adults who have First Amendment rights to access a broader range of materials on the Internet than are allowed under the Children's Internet Protection Act. One county library was sued for filtering while another was sued for not filtering enough. At least one federal court has ruled that a local library board may not require the use of filtering software on all library Internet computer terminals. A possible compromise for public libraries with multiple computers would be to allow unrestricted Internet use for adults, but to provide computers with only limited access for children. Rather than deal with all the technical and legal complications, some librarians say they wish they could simply train students and adults to use the Internet safely and wisely. After all, bad behavior in libraries did not start and will not end with the Internet.

The ACLU filed a suit to challenge the Children's Internet Protection Act. In May 2002, a three-judge panel declared the act unconstitutional. In their ruling, the judges agreed with arguments made by the ACLU and others that blocking programs cannot effectively screen out only material deemed "harmful to minors."[4] The court said too much erroneous filtering caused by the law resulted in violation of the First Amendment. "It is currently impossible, given the Internet's size, rate of growth, rate of change and architecture ... to develop a filter that neither underblocks nor overblocks a substantial amount of speech," said the justices.[5]

An appeal of this decision is expected and the case will then go directly to the United States Supreme Court, which is required to hear challenges to this law.

Anonymity

Anonymous expression allows one to state one's opinions without revealing one's identity. The freedom to express an opinion without fear of reprisal is an important right of a democratic society. Anonymity is even more important in countries that don't allow free speech. However, in the wrong hands, anonymous communication can be used as a tool to commit illegal or unethical activities.

Anonymous political expression played an important role in the early formation of the United States. Before and during the American Revolution, patriots who dissented against British rule often used anonymous pamphlets and leaflets to express their opinions. England had a variety of laws designed to restrict anonymous political commentary and those found guilty of breaking those laws were subject to harsh punishment—from whippings to hangings. A famous case in 1735 involved a printer named John Zenger, prosecuted for seditious libel because he wouldn't reveal the names of anonymous authors whose writings he published. The authors were critical of the governor of New York. The British were outraged when the jurors refused to convict Zenger in what is considered one of the defining moments in the history of freedom of the press.

Thomas Paine was an influential writer, philosopher, and statesman of the American Revolutionary War era. He published a pamphlet called "Common Sense," in which he criticized the British monarchy and urged the Colonies to become independent and to establish a republican government of their own. Published anonymously in 1776, the pamphlet sold more than 500,000 copies and provided a stimulus to produce the Declaration of Independence six months later. Other pro-democracy supporters often authored their writings anonymously or under pseudonyms. Thus, looking back at the formative years of America, one can see that anonymous political expression was widely practiced.

Despite the importance of anonymity in early America, it took nearly 200 years for the Supreme Court to render rulings that address anonymity as an aspect of the Bill of Rights. One of the first rulings was in the case of *NAACP vs. Alabama* in 1958, where the court ruled that the NAACP did not have to turn over its membership list to the state of Alabama. The court believed that members could be subjected to threats and retaliation if the list were disclosed. Disclosure would restrict a member's right to freely associate in violation of the First Amendment.

Yet another landmark anonymity case involved a sailor threatened with discharge from the United States Navy because of information obtained from America Online. In

1998, following a tip, a Navy investigator asked America Online for the real identity behind the sailor's Internet pseudonym. Under that name, the sailor had posted information in a personal profile that suggested he might be gay. Thus, he could be discharged under the military's "don't ask, don't tell" policy on homosexuality. America Online admitted that their representative violated company policy in providing the information. A federal judge ruled that the Navy had overstepped its authority in probing the sailor's sexual orientation and had also violated the Electronic Communications Privacy Act, which limits how government agencies can seek information from e-mail or other online data. The sailor received undisclosed monetary damages from America Online and, in a separate agreement, was allowed to retire from the Navy with full pension and benefits.[6]

Maintaining anonymity on the Internet is important to some Internet users. Such users might be seeking help in an online support group, reporting defects about a manufacturer's goods or services, participating in frank discussions of sensitive topics, expressing a minority or anti-government opinion in a hostile political environment, or participating in chat rooms. Other Internet users would like to ban anonymity because they think that its use increases the risks of defamation, fraud, libel, and exploitation of children by enabling people to conceal their identities.

There are companies that specialize in investigations to identify anonymous individuals engaged in illegal activities. For example, Internet Crimes Group, Inc. was used to find a person using an Internet pseudonym who had posted several false allegations concerning a company's officers and financial results. In the course of the investigation, nearly two-dozen additional screen names were revealed. It was discovered that the anonymous poster was a National Association of Securities Dealers licensed broker. The broker was conducting these illegal activities to enhance an investment portfolio under his management. The case was referred to the Securities and Exchange Commission (SEC) Internet Enforcement Division for follow-up.[7]

Anonymous Remailers

Maintaining anonymity is a legitimate need for some Internet activities; however, the address in an e-mail message or Usenet newsgroup posting clearly identifies its author. Internet users wishing to remain anonymous can send e-mail to an **anonymous remailer** service, where a computer program strips the originating address from the message. It then forwards the message to its intended recipient—an individual, chat room, or newsgroup—with either no address or a fictitious one. This ensures that there is no header information that can identify the author. Some remailers use encryption and routing through multiple remailers to provide a virtually untraceable level of anonymity.

The use of a remailer keeps one's communications anonymous; what is communicated, and whether it is ethical, is up to the person sending the communication. The use of remailers to enable people to do things considered unethical by others or even illegal in some states or countries has spurred controversy. Remailers are frequently used to send out pornography, to illegally post copyrighted material to Usenet newsgroups, and to send unsolicited advertising to broad audiences (spamming). However, remailers also meet e-mailers' legitimate needs for maintaining anonymity. A corporate IT organization may wish to employ filters or set the corporate firewall to prohibit employees from accessing remailers, or to send a warning message each time an employee communicates with a remailer.

John Doe Lawsuits

Business organizations must be on the alert to protect against the expression of opinions detrimental to their reputations (as illustrated in the Intel example at the start of this chapter) or the public sharing of company confidential information. When such attacks are launched via anonymous individuals over the Internet, the potential for widescale sharing of harmful information is enormous and inspires great effort to identify the individuals involved and put a stop to their actions.

A **John Doe lawsuit** is one for which the true identity of the defendant is temporarily unknown. Such suits are common in Internet libel cases, where the defendant is unknown because he or she communicates using a pseudonym or anonymously. Often the parties filing such lawsuits are corporations upset by the unknown defendant's e-mail messages that are critical of the company or that reveal company secrets. For example, Raytheon filed a lawsuit in 1999 for $25,000 in damages against 21 "John Does" for allegedly revealing on a Yahoo! message board Raytheon financial results and other information that the company claimed hurt its reputation. Pursuant to the suit, Raytheon requested and received a court order to subpoena Yahoo and several ISPs for the identity of the 21 unnamed defendants. Eventually, Raytheon was able to trace the identities of all 21 who posted the alleged company secrets. Four employees voluntarily left the company and the others involved received counseling about sharing confidential company information.[8]

America Online, Yahoo!, and other ISPs receive nearly a thousand subpoenas a year to reveal the identity of John Does. Free speech advocates argue that if someone charges libel, then the anonymity of the poster should be preserved until the libel is proved. Otherwise, the subpoena power can be used to silence anonymous, critical speech.[9]

Proponents of such lawsuits point out that most John Doe lawsuits are based on serious allegations of wrongdoing, such as libel or disclosure of confidential information. For example, stock-price manipulators can use chat rooms to affect the share price of a stock (especially those of very small companies with just a few outstanding shares). In addition, competitors of an organization might try to create a feeling that a company is a miserable place to work to discourage job candidates from applying, investors from buying stock, or consumers from buying its products. John Doe lawsuit proponents argue that the perpetrators should not be able to hide behind anonymity to avoid responsibility for their actions.[10]

Anonymity on the Internet is not guaranteed. By filing a lawsuit, companies gain immediate subpoena power, and many message board hosts release information right away, often without notifying the poster. All who post comments in a public place on the Internet must bear in mind what the consequences might be if their identities are exposed. Furthermore, everyone who reads anonymous postings on the Internet should think twice about believing what they read. Read the Legal Overview to learn more about the court precedents that have been set with regard to the issue of protecting the identity of anonymous Internet posters.

LEGAL OVERVIEW

SAFEGUARDING THE IDENTITY OF ANONYMOUS INTERNET POSTERS

The California State Court ruling in *Pre-Paid Legal v. Sturtz et al* set a legal precedent that courts will apply to subpoenas requesting the identity of anonymous Internet speakers. The case involved a subpoena issued by Pre-Paid Legal Services (PPLS) requesting the identity of eight anonymous posters on Yahoo!'s "Pre-Paid" message board.

Attorneys for PPLS argued that it needed the posters' identities to determine whether they were subject to a voluntary injunction preventing former sales associates who work for a competitor from revealing PPLS's trade secrets.

The EFF represented two of the John Does whose identities were subpoenaed. EFF attorneys argued that the message board postings cited by PPLS revealed no company secrets. They merely indicated that the eight John Does were disparaging the company and its treatment of its sales associates. They argued furthermore that requiring the John Does to reveal their identities would let the company punish them for speaking out and set a dangerous precedent that would discourage other Internet users from voicing criticism. Without proper safeguards on John Doe subpoenas, a company could use the courts to uncover its critics.

EFF attorneys urged that the court apply a four-part test adopted by the federal courts in *Doe v. 2TheMart.com, Inc.* to determine whether a subpoena for the identity of Internet speakers should be upheld. In that case, the federal court ruled that a subpoena asking an Internet company to reveal the identities of persons who use its service to speak anonymously on the Internet should be enforced only when the subpoena was 1) issued in good faith and not for any improper purpose, 2) the information sought related to a core claim or defense, 3) the identifying information was directly and materially relevant to that claim or defense, and 4) adequate information was unavailable from any other source.

In August 2001, a judge in the Santa Clara County Superior Court invalidated the subpoena to Yahoo! requesting the posters' identities. His ruling was that the messages were not obvious violations of the injunctions invoked by PPLS, and therefore, the First Amendment protection of anonymous speech outweighed PPLS's interest in learning the identity of the speakers.

Sources: adapted from John Soat, "IT Confidential," *Information Week*, August 20, 2001, accessed at www.informationweek.com; John Rendleman, "Praise for Court's Privacy Decision," *Information Week*, August 14, 2001, accessed at www.informationweek.com; "Court Protects Online Anonymity of Corporate Critics," Electronic Frontier Foundation Media Release, August 13, 2001, accessed at www.eff.org.; "EFF Asks Court to Quash John Doe Subponea," Electronic Frontier Foundation Media Release, August 7, 2001, accessed at www.eff.org.; "Free Speech Advocates Seek to Protect Anonymous Speech on Internet," Electronic Frontier Foundation Press Release, February 26, 2001, accessed at www.eff.org.

Defamation and Hate Speech

In the United States, "speech" that is merely annoying, critical, demeaning, or offensive enjoys protection under the First Amendment. Legal recourse is possible only when hate speech turns into clear threats and intimidation against specific individuals. Persistent or malicious harassment aimed at a specific individual can be prosecuted under the law. General, broad scope statements expressing hatred of an ethnic, racial, or religious group cannot. A threatening private message sent over the Internet to an individual, a public message displayed on a Web site describing intent to commit acts of hate-motivated violence, and libel directed at a particular person are all actions that can be prosecuted.

In addition to First Amendment protection, another difficulty in enforcing laws against defamation and hate speech is the ease of anonymous communication over the Internet. Anyone can send hateful e-mails and avoid easy identification by using a remailer service that guarantees complete anonymity. America Online and other ISPs have, from time to time, voluntarily agreed to prohibit their subscribers from sending hate messages using their services. Because such prohibitions are included in the service contracts between a private company (the ISP) and its subscribers and do not involve the federal government, they do not violate the subscribers' First Amendment rights.

After an ISP implements such a prohibition, it must monitor the use of its service to ensure that the regulations are followed. When a violation occurs, the ISP must take action to prevent it from happening again. For example, if a subscriber who is a participant in a chat room engages in hate speech in violation of the terms of service of the ISP, the subscriber's account can be cancelled or the subscriber can be warned and forbidden from participating in the chat room. To aid in the enforcement of such prohibitions, the ISP can also encourage subscribers to report violators to the appropriate company representatives. Of course, ISP subscribers who lose an account for violating that ISP's regulations may resume their hate speech by simply opening a new account with some other, more permissive ISP.

Public schools and universities are legally considered to be agents of the government and therefore must follow the First Amendment's prohibition against speech restrictions based on content or viewpoint. Corporations, private schools, and universities, on the other hand, are not part of the state or federal government. As a result, they may prohibit students, instructors, and other employees from engaging in offensive speech using corporate, school, or university computers, Internet, or e-mail services.

Despite the protection of the First Amendment and the challenges posed by anonymous expression, there are instances of United States citizens being successfully sued or convicted of crimes relating to hate speech:

- In 1998, a former student was sentenced to one year in prison for sending e-mail death threats to Asian-American students at the University of California, Irvine. His e-mail was signed "Asian hater," and in his letters, he said he would make it his life career to find and kill every Asian personally.
- In 1999, a coalition of groups opposed to abortion was ordered to pay more than $100 million in damages. The group had placed information on a Web site that included photos of doctors and clinic workers who

perform abortions, their home addresses, license plate numbers, and even the names of their spouses and children. Three of the doctors listed on the site were murdered and others were wounded. A jury found that the Web site provided information that resulted in a real threat of bodily harm and awarded damages. However, in March 2001, the United States Court of Appeals for the Ninth Circuit reversed this decision. The court ruled that the coalition of groups made no statements mentioning violence and that publication of the personal information did not constitute a serious expression of intent to harm. [11]

- In 2001, Varian Medical Systems won a $775,000 jury verdict in an Internet defamation and harassment lawsuit against former employees who posted thousands of messages that accused managers of being homophobic and of discriminating against pregnant women.[12]

Most other countries do not provide constitutional protection for hate speech. Promoting Nazi ideology is a crime in Germany. Denying the occurrence of the Holocaust is illegal in many European countries. Authorities in Britain, Canada, Denmark, France, and Germany have charged people for crimes involving hate speech on the Internet.

A United States citizen who posts material on the Internet that is illegal in a foreign country can be prosecuted if he subjects himself to the jurisdiction of that country by, for example, visiting there. As long as the individual remains in the United States, he or she is safe from prosecution, as our laws do not allow a person to be extradited for engaging in an activity protected by the United States Constitution, even if that activity violates the criminal laws of another country.

Pornography

Many adults and free speech advocacy groups believe that there is nothing illegal or wrong about purchasing adult pornographic material made for and by consenting adults. They argue that the First Amendment protects such material. On the other hand, most parents, educators, and other child advocates are upset by the thought of children viewing such material. They are deeply concerned about the impact of pornography on children and fear that the increasingly easy access to pornography encourages pedophiles and sexual molesters.

Clearly the Internet has been a boon to the pornography industry by providing fast, cheap, and convenient access to over 60,000 Web sex sites. Perhaps most importantly, access via the Internet enables pornography consumers to avoid offending others or being embarrassed by others observing their purchases (such as might occur during the rental of pornographic tapes in a local video store). There is no question that adult pornography on the Internet is big business and generates lots of traffic. About one in four regular Internet users (almost 21 million Americans) visit a Web sex site at least once a month—more than the number of visitors to Web sports sites. Forrester Research estimates that sex sites on the Web generate at least $1 billion a year in revenue.[13]

United States organizations must exercise great care in how they deal with the issue of pornography in the work place. By providing employees with computers and access to the Internet, organizations have enabled their employees to access this material. It is against United States obscenity laws to publish pornography, and companies can be seen in the eyes of the law to be publishers because they have provided the tools, technology, and training to enable employees to store files of pornographic material on their hard disks and retrieve it on demand.

Some companies believe that they have a duty to stop the viewing of pornography in the workplace. As long as they can show that they were taking reasonable steps to prevent it, they have a valid defense if they were ever the subject of a sexual harassment lawsuit. However, they must be able to show that they have taken determined actions to do this. If it can be shown that they made just a half-hearted attempt at stopping it but gave up as soon as things got a bit difficult, then their defense in court would be weak. This means establishing a computer usage policy that prohibits access to pornography sites, identifying those who violate the policy, and taking action against those individuals—no matter how embarrassing it is for the individuals or harmful it might be for the company.

A few companies take an opposite viewpoint. Their legal and human resources people believe that the company cannot be held liable if they don't know employees are viewing, downloading, and distributing pornography. Thus, they advise management to ignore the problem by never investigating whether there is a problem, thereby ensuring that they can claim that the company never knew it was happening. Many people would consider such an approach unethical and would view this as management shirking an important responsibility to provide a work environment free of sexual harassment. Employees unwillingly exposed to such material would indeed have a strong case for sexual harassment because they could claim that pornographic material was available in the workplace and the company took inadequate measures to control the situation.

In contrast to adult pornography, there are numerous federal laws addressing child pornography. As a result of these laws, possession of child pornography is a federal offense punishable by up to five years in prison. The production and distribution of such materials carry harsher penalties—decades or even life in prison is not an unusual sentence. In addition to these federal statutes, all states have enacted laws against the production and distribution of child pornography. All but a few states have outlawed the possession of child pornography. South Carolina went so far as to pass a law in July 2001 that requires computer technicians who view child pornography when working on a computer to give the names and addresses of the PC user or owner to law enforcement officials.[14]

See Table 5-1 for a manager's checklist for dealing with issues of freedom of expression in the workplace. In each case, the preferred answer is "Yes."

Table 5-1   Manager's checklist for dealing with freedom of expression issues at work

Questions	Yes	No
Does your corporate IT usage policy discuss the need to conserve corporate network capacity, avoid legal liability, and improve worker productivity by limiting the non-business use of information resources?	___	___
Have means been implemented to limit employee access to non-business oriented Web sites (e.g. Internet filters, firewall configurations, or use of a ISP that block access to such sites)?	___	___
Does your corporate IT usage policy discuss the inappropriate use of anonymous remailers?	___	___
Has your corporate firewall been set to detect the use of anonymous remailers?	___	___
Has your company (in cooperation with legal counsel) formed a policy on the use of John Doe lawsuits to identify the authors of libelous, anonymous e-mail?	___	___
Does your corporate IT usage policy make it clear that defamation and hate speech have no place in the business setting?	___	___
Does your corporate IT usage policy prohibit the viewing or sending of pornography?	___	___
Is employee e-mail regularly monitored for the sending or receiving of defamation, hate mail, and/or pornography?	___	___
Does your corporate IT usage policy tell employees what to do in the event that they receive hate mail or pornography?	___	___

Summary

1. What is the legal basis for the protection of our freedom of speech and what types of speech are not protected under the law?

    The First Amendment protects the right to freedom of religion and freedom of expression from government interference. Numerous court decisions have broadened the definition of speech to include non-verbal, visual, and symbolic forms of expressions. Sometimes the "speech" at issue is unpopular or highly offensive to the majority of people; however, the Bill of Rights provides protection for the views of the minority. The Supreme Court has also ruled that the First Amendment protects the right to speak anonymously as part of the guarantee of free speech. The following types of speech are not protected by the First Amendment and may be totally forbidden by the government: obscene speech, defamation, incitement of panic, incitement to crime, "fighting words," and sedition.

2. In what ways does the Internet present new challenges in the area of freedom of expression?

    The Internet enables a worldwide exchange of news, ideas, opinions, rumors, and information. Its broad accessibility, open-minded discussions, and anonymity make it an ideal communications medium. Individuals must often make ethical decisions in regard to how they will use such remarkable freedom and power. Organizations and governments have attempted to establish policies and laws to help guide individuals as well as protect their own interests. Business organizations, in particular, have sought to conserve corporate network capacity, avoid legal liability, and improve worker productivity by limiting the non-business use of information technology resources.

3. What are the key freedom of speech issues that must be addressed in considering the use of information technology?

    Controlling access to Internet information is one key issue. Although there are clear and convincing arguments to support freedom of speech on the Internet, the issue becomes problematic when one considers the ease with which children can access the Internet. The conundrum is that it is difficult to restrict Internet access by children without at the same time restricting access by adults. The United States government has passed several laws in an attempt to address this issue, and software manufacturers have invented special software whose goal is to block access to objectionable material.

    Anonymous communication is another key issue. Business organizations must be on the alert to protect against the expression of opinions detrimental to their reputation or the public sharing of company confidential information. When such attacks are launched via anonymous individuals over the Internet, the potential for widescale sharing of harmful information is enormous. Businesses go to great lengths to identify the individuals involved and put a stop to their actions. Internet users wishing to remain anonymous can send e-mail to an anonymous remailer service, but whether what is communicated is ethical is up to the person sending the e-mail.

    The spread of defamation and hate speech is another key issue, especially for ISPs. In the United States, Internet "speech" that is merely annoying, critical, demeaning, or offensive enjoys protection under the First Amendment. Legal recourse is possible only when hate speech turns into clear threats and intimidation against specific individuals. ISPs have, from time to time, voluntarily agreed to prohibit their subscribers from sending hate messages using their services. Because such prohibitions can be included in the service contracts between a private company (the ISP) and its subscribers and do not involve the federal government, they do not violate the subscribers' First Amendment rights. After an ISP implements such a prohibition, it must monitor the use of its service to ensure that the regulations are followed. When a violation occurs, the ISP must take action to prevent it from happening again.

The use of information technology to access, store, and distribute pornography in the workplace is another key issue. Organizations must exercise great care in how they deal with this issue. As long as they can show that they were taking reasonable steps to prevent it, they have a valid defense if they are subject to a sexual harassment lawsuit. This means establishing a computer usage policy that prohibits access to pornography sites, identifying those who violate the policy, and taking action against those individuals—no matter how embarrassing it is for the individuals or harmful it might be for the company.

Review Questions

1. What is the First Amendment? To whom does it apply?
2. What is obscene speech? Is it protected under the First Amendment?
3. What was the Communications Decency Act (CDA)? Why was it ruled unconstitutional?
4. Discuss the role that anonymous expression played in the early formation of the United States.
5. What are some actions you can take if you wish to remain anonymous on the Internet? Are there ways in which your identity can be uncovered?
6. Give several examples of speech that is not protected by the First Amendment.
7. Why can corporations and private universities prohibit students, instructors, and other employees from engaging in offensive speech using company equipment or company services while public universities cannot?
8. What are the two extreme viewpoints taken by companies when dealing with pornography in the workplace?

Discussion Questions

1. Outline a scenario in which you might be acting ethically but still want to remain anonymous while using the Internet.
2. Do you think exceptions should be made to the First Amendment to make it easier to prosecute individuals who send hate e-mails and use the Internet to try to recruit people to their cause? Why or why not?
3. What actions can an ISP take to limit the distribution of hate e-mail and discussion? Why are such actions not considered a violation of the subscriber's First Amendment rights?
4. Briefly describe how the ICRA Web site rating process works. What are the advantages and disadvantages of this system?
5. What is a John Doe lawsuit? Do you think that corporations should be allowed to use a subpoena to identify a John Doe before proving that the individual has defamed the company? Why or why not?
6. Draft a brief description of a new Internet law designed at protecting children from being exposed to indecent material on the Internet. Avoid the basic problems with the CDA and the Children's Internet Protection Act. Write another paragraph pointing out the strengths and weaknesses of your proposed law.

What Would You Do?

1. You are the vice-president of human resources and are working with a committee to complete your company's computer usage policy. How would you advise the committee about creating a corporate policy on Internet pornography—would the policy be laissez-faire or would it require strict enforcement of tough corporate guidelines? Why?

2. You are a member of the computer support group of your company and have just made a visit to a user to upgrade his computer. As you are testing things out after making the upgrade, you are surprised to find that the user has "disabled" the Internet filter software that is supposed to be standard on all corporate computers. What would you do?

3. Imagine that at your school or place of work you receive a hate e-mail. Does your school or place of work have a policy that covers such issues? What would you do?

4. You are the information systems manager of your county's public library system. You purchased 150 licenses of filtering software in anticipation of installing it when the Children's Internet Protection Act was first passed. Several members of the library's board of directors want the filtering software installed on computers designated for use by children only. The board is evenly divided on this matter 3–3, with 1 abstention. They have asked you to present your recommendation at the next meeting.

Supporters of implementing the filters say that filter technology is improving and will be able to provide adequate protection without being overly restrictive. Innocent children need to be shielded from objectionable material and the filters should be installed and perfected over time.

Opponents of installing the filters say there is no such thing as a perfect filter. They argue that if such software blocks access to both a Jackie Collins novel and obscene material because it can't tell the difference, it cannot be used to police what individuals view at public libraries.

What would you say to the board?

Cases

1. The Electronic Frontier Foundation

The EFF was founded in 1990 by John Perry Barlow (lyricist for the Grateful Dead) and Mitch Kapor (founder of the Lotus 1-2-3 spreadsheet software company). It is a nonprofit, non-partisan organization whose goal is to protect fundamental civil liberties related to technology, including privacy and freedom of expression on the Internet. It frequently undertakes court cases as an advocate of the preservation of individual rights.

The EFF's mission includes the education of the press, policymakers, and the general public about civil liberties issues. To this end, its Web site (www.eff.org) provides an extensive collection of information on issues such as censorship, free expression, digital surveillance, encryption, and privacy issues. The Web site gets more than 100,000 hits a day and is among the most visited sites on the Internet. Many of the visitors are key managers within companies responsible for making decisions about how to act ethically and legally in applying information technology.

Although over 80 percent of its annual operating budget comes from individual donations from concerned citizens, there are also a number of corporate sponsors, including the Boston Consulting Group, Intuit, Netscape, Oracle, Pacific Bell, Sun Microsystems, Tandem Computer, the law firm of Wilson, Sonsini, Goodrich, Rosati (specializing in intellectual property issues), the law firm of Latham Watkins (a corporate finance and business litigation firm), the investment banking firms of Goldman Sachs and Morgan Stanley, and the public relations firm of Fleischmann-Hilliard. Many of these sponsors hope to do a better job of minimizing the negative impacts of information technology on society and help information technology users become more responsible and proactive in fulfilling their role and responsibilities. If this happens, it is good for business because it builds a healthier and more knowledgeable marketplace in which to sell their products. On the other hand, the EFF has developed many critics over the years for what some see as its bias against most forms of regulation.

Questions:

1. What reasons might a firm give for joining and supporting the EFF?
2. Visit the EFF Web site and develop a list of the current "hot issues." Research one EFF issue of interest to you. Write a brief paper summarizing the EFF's position. Discuss whether you support this position and why.
3. The vice president of public affairs for your mid-sized telecommunications equipment manufacturing company has suggested that the firm donate $100,000 to the EFF and become a corporate sponsor. The CFO has asked if you, the CIO, support this action. What would you do?

Sources: adapted from Peter Kizilos, "(interview with) Lori Fena," *Online*, March–April 1998, accessed at www.findarticles.com and the EFF Web site at www.eff.org, accessed on June 18, 2002.

2. SurfControl

SurfControl develops software products that filter both Internet and e-mail use. The firm's original target market was the corporate world. Its goal was to provide products that encouraged responsible Internet usage by reporting how the Internet was being used. Thus, businesses could implement and enforce Internet access policies. Since going public in June of 1998, SurfControl has completed six acquisitions, including two of the early pioneers in the home filtering market—SurfWatch and CyberPatrol. As a result, SurfControl now has products that address both Internet and e-mail filtering needs, and it has expanded its market base to include corporate, education, home, and technology OEM (original equipment manufacturer) customers.

The SuperScout Web Filter is SurfControl's flagship business-oriented Web filtering product. Businesses employ the SuperScout product to protect against network resource hogs and time diverters such as personal banking, non-business-related travel planning, and the downloading of music. SuperScout reports who is going where and when on the Internet, and it offers the ability to define detailed rules for Internet filtering—down to the employee level and the Web-page level. It includes a feature that enables users to manage their own surfing within pre-set time or data transfer limits. SurfControl is currently developing an image recognition engine that will further enhance its Internet filtering capabilities.

SuperScout E-mail Filter is used by businesses to stop viruses from infecting their corporate network by providing anti-virus scanning of e-mails at the corporate gateway to the Internet. The software can also protect employees from receiving spam and other junk e-mail that undermines productive work. On the outgoing e-mail side, the software can prevent the loss of confidential information by blocking, isolating, or delaying any e-mail containing intellectual property or other confidential information until it has been approved. Encrypted messages sent by unauthorized personnel within a company can also be blocked. All non-essential e-mail can be scheduled to be transmitted outside prime work hours, ensuring priority is given to business critical e-mail during the workday.

The Electronic Communications Privacy Act (ECPA) prohibits the interception of any wire, oral, or electronic communication, or the unauthorized access of stored communications. There are, however, three key exceptions, and if any one of these applies, monitoring can occur without violation of the act.

Employers are allowed to 1) monitor business-related phone calls, 2) monitor communications when there has been employee consent, and 3) retrieve and access stored e-mail messages. To avoid legal entanglements, employers should work with legal counsel to develop business equipment usage policies that describe how they intend to monitor employees.

Questions:

1. Which features of the SuperScout Web and E-Mail Filter products do you consider to be the most worthwhile? Which features are not so attractive? Why?

2. What sort of consulting services would you expect to see offered in addition to help with the installation of this software?

3. What actions would you recommend before using such software to monitor employee Web-surfing habits and e-mail?

Sources: adapted from SurfControl Web site at www.surfcontrol.com, accessed on June 17, 2002; "Massachusetts Internet Filtering Technology Company Says Mandatory Filtering Laws Aren't Needed," *PRNewswire*, June 4, 2001, accessed at www.prnewswire.com; Eric J. Sinrod, "Who's Watching You?", *Computerworld*, June 11, 2001, accessed at www.computerworld.com.

Endnotes

[1] "Supreme Court Rules: Cyberspace Will be Free! ACLU Hails Victory in Internet Censorship Challenge," June 27, 1996, American Civil Liberties Union Web site at www.aclu.org, accessed on August 30, 2001.

[2] Leslie Jaye Goff, "Technology Flashback—1996 Internet Content Decency Debate," *Computerworld*, November 29, 1999, accessed at www.computerworld.com.

[3] "U.S. Supreme Court Decision, Reno, et al v. ACLU, et al (June 27, 1997)" accessed at www2.epic.org on August 31, 2001.

[4] "Federal Court Rejects Government Censorship in Libraries, Citing Free Speech Rights of Patrons," May 31, 2002, ACLU Web site at www.aclu.org.

[5] Anne Ju, "Can The 'Net Ever Be Made Safe for Kids?" *Computerworld*, June 6, 2002, accessed at www.computerworld.com.

[6] Sharon Machlis, "Sailor Settles with AOL, Navy," *Computerworld*, June 12, 1998, accessed at www.computerworld.com.

[7] Internet Crimes Group, Inc. home page, accessed at www.internetcrimesgroup.com on August 21, 2001.

[8] Kaitlin Quistgaard, "Raytheon Triumphs over Yahoo Posters' Anonymity," *Technology*, May 24, 1999, accessed at www.salon.com.

[9] John Schwartz, "Questions on Net Anonymity," *The New York Times on the Web*, October 17, 2000, accessed at www.nytimes.com.

[10] Carl S. Kaplan, "Virginia Court's Decision in Online 'John Doe' Case Hailed by Free-Speech Advocates," *The New York Times on the Web*, March 16, 2001, accessed at www.nytimes.com.

[11] "Planned Parenthood of the Columbia/Willamette Inc. v. American Coalition of Life Activists (March 28, 2001)," accessed at www.ce9.uscourts.gov on August 31, 2001.

[12] "Companies' Lawsuits Seek to Silence Employees," *ACLU News*, January 7, 2002, accessed at the ACLU Web site at www.aclu.org.

[13] Timothy Egan, "Wall Street Meets Pornography," *The New York Times on the Web*, October 25, 2000, accessed at www.nytimes.com.

[14] Sandra Swanson, "You Be The Judge," *Information Week*, August 6, 2001, pp. 18–20.

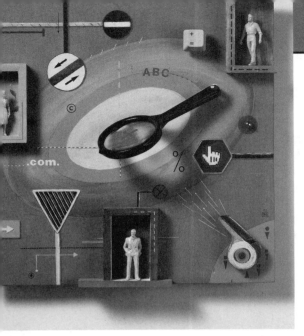

Chapter 6

Intellectual PROPERTY

"The fundamental problem is that copyright pretends that information is property. The Internet is the most effective communications technology we've ever had. It's inevitable that it's going to make it difficult to enforce copyright."
— Ian Clarke, developer of the Freenet platform that can be used to swap files online[1]

In August 2001, a grand jury indicted ElcomSoft, a Moscow-based company, and one of its employees for selling software designed to circumvent the Digital Millennium Copyright Act (DMCA). This United States law bans the sale of technology that can allow people to overcome software, encryption, or other mechanisms that protect copyrighted material in digital programs. ElcomSoft briefly sold a computer program over the Internet that enabled people using Adobe Systems' eBook Reader to circumvent copyright protections so that they could copy and print digital books, as well as transfer them to other computers and have the computer read them aloud. Under the charges, the employee faced up to 25 years in prison and a fine of up to $2.25 million, and ElcomSoft, as a corporation, faced a penalty of $2.5 million.

ElcomSoft said the program lets legitimate buyers of e-books stretch the limits imposed by Adobe's software, which keeps them from making a copy of a book to read on an additional computer or to use as a backup in case it is erased. The company also maintains that the program is legal in

Russia and that fewer than 10 copies were sold before its sale was stopped under legal pressure from Adobe Systems.

Some industry observers say that efforts to police the security of e-books may ultimately work against the hopeful e-book market, which unlike the digital music and video industries, has yet to take off. The electronic book industry is a small but competitive industry in which the key players are Adobe, Gemstar, Microsoft, and Palm.

The indictment states that the program was made available for download from a Web server located in Chicago and customers were directed to make payment through an online payment service based in Washington state. As a result, legally, sales of the product took place partly in the United States. In spite of the borderless nature of the Internet, the fact that the case involves a Russian citizen and a Russia-based company, and the fact that the DMCA carries no weight outside the United States, the question of jurisdiction was not contested.

In December 2001, the programmer who wrote the software was released with the promise that charges would be dropped against him in exchange for his testimony. As of this writing, a trial date had not been set.

Sources: Jennifer Lee, "U.S. Arrests Russian Cryptographer as Copyright Violator," *The New York Times on the Web*, July 18, 2001, accessed at www.nytimes.com; Reuters, "Russian Programmer and Employer Indicted," *The New York Times on the Web*, August 29, 2001, accessed at www.nytimes.com; "Judge Rules Russia Software Firm Can Be Charged on Copyright Charge," *San Jose Mercury News*, May 8, 2002, accessed at www.siliconvalley.com.

As you read this chapter, consider the following questions:

1. What does the term "intellectual property" encompass and why are companies so concerned about protecting it?

2. What are the strengths and limitations of using copyrights, patents, and trade secret laws to protect intellectual property?

3. What is reverse engineering and what are the issues associated with applying this technique to uncover the trade secrets of a competitor's software program?

4. What is the purpose of the Uniform Computer Information Transactions Act and what is its potential future impact on software licensing?

5. What is the essential difference between competitive intelligence and industrial espionage and what are several sources of competitive intelligence information?

6. What strategy should be used to protect an organization from cyber-squatting?

WHAT IS INTELLECTUAL PROPERTY?

Intellectual property is a term used to describe works of the mind such as art, books, films, formulae, inventions, music, and processes that are distinct and that are "owned" or created by a single entity.

Copyright law protects authored works such as art, books, film, and music. Patent laws protect inventions. Trade secret protection laws help enforce the concepts of fairness in business dealings. Together copyright, patent, and trade secret laws form an intricate body of law addressing the ownership of intellectual property. Intellectual property is relevant to the discussion of ethics in IT because this body of law offers both hope and potential ethical problems for the creators and users of information and information technology. Because a prodigious amount of copyrighted or patented information is or is capable of being stored electronically, it is easy to provide access to such information to those who need or want to use it. Every day, more and more data is stored in a digital form that allows for flawless, easy copying of those files to other people's computers. There are many potential problems with providing such access, though—information can easily be reproduced inappropriately, at the one extreme, and, at the other extreme, access to information can be limited in new and challenging ways. Many skeptics believe that laws, copyrights, patents, and trademarks are stifling creativity by making it harder to build on the ideas of others.

Defining the appropriate level of access to intellectual property and then maintaining control over access to that intellectual property are complex tasks. For example, protecting computer software has proven to be extremely difficult because it is unclear how software programs should be categorized under the law. Software has sometimes been treated as the expression of an idea that can be protected under copyright law. Software also has been treated as a process for changing the internal structure of a computer and thus eligible for protection under patent law. In some cases, software has been judged to be a series of mental steps and, thereby, not appropriate for ownership and ineligible for any form of protection.

Copyrights

Copyright and patent protection is established in the United States Constitution in Article I, section 8, clause 8, where it is specified that Congress shall have the power "To promote the Progress of Science and useful Arts, by securing for limited Times to Authors and Inventors the exclusive Rights to their respective Writings and Discoveries."

A copyright is a form of protection provided by the United States Code, which is the general and permanent laws of the United States. A **copyright** grants the authors of "original works of authorship in any tangible medium of expression, now known or later developed, from which they can be perceived, reproduced, or otherwise communicated, either directly or with the aid of a machine or device, the exclusive right to distribute, display, perform, or reproduce the work in copies or to prepare derivative works based upon the work." (United States Code, Title 17, Section 102(a)) The author may grant this exclusive right to others. As new forms of expression develop, they can be awarded copyright protection. For example, audiovisual works were added and computer programs were assigned to the literary works category with the Copyright Act of 1976.

Copyright law guarantees to the developers of these works the rights to their works, but for limited times. For example, Section 106 of the 1976 Copyright Act gives the owner of the copyright the right to prevent others from reproducing the work for the author's life plus 70 years. If the author produces the work while in the employment of another, it is the employer who is considered to be the author and a copyright for such works lasts 95 years. Since 1960, the term of copyright, originally 28 years, has been extended eleven times, most recently from 75 to 95 years. This extension has been championed mainly by motion picture studios that wish to retain the rights to their early films. Opponents of the extension argue that the lengthening of the copyright period makes it more difficult for current artists to build on the work of others, thus stifling creativity and innovation.

The types of work that can be copyrighted include architecture, art, audiovisual works, choreography, drama, graphics, literature, motion pictures, music, pantomimes, pictures, and sculptures, sound recordings, and other intellectual works as described in U.S.C. Title 17. To be eligible for a copyright, a work must fall within one of the preceding categories, and it must also be original. Copyright law has proven to be extremely flexible in covering new technologies: software, videogames, multimedia works, and Web pages can all be protected by copyright. However, evaluation of the originality of a work can be problematic and has given rise to much litigation.

Works that have not been fixed in a tangible form of expression (such as an improvisational speech that has not been written down or recorded) and work consisting entirely of information that is common property and that contains no original authorship (such as a chart showing the conversion between metric and American units of measure) are not eligible for copyright protection.

Copyright law tries to strike a balance between protecting the author's rights and enabling the public to access copyrighted works. To this end, the **fair use doctrine** (U.S.C., Title 17, section 107) sets forth four factors for courts to consider in determining whether a particular use of copyrighted property is a fair use and can be allowed without penalty. These four factors are 1) the purpose and character of the use (such as whether the use is for commercial use or for nonprofit, educational purposes); 2) the nature of the copyrighted work; 3) the portion of the copyrighted work used; and 4) the effect of the use upon the value of the copyrighted work.

The concept that an idea cannot be copyrighted, but the expression of an idea can be, is a key to understanding copyright protection. For example, an author writing about the history of the United States cannot copy the exact words used by another to describe the feelings of a participant in a battle from World War II, but he or she can

135

still convey the sense of terror and horror felt by that individual. Another key aspect of copyright is that if someone else independently (without knowledge of a pre-existing work) develops something similar or even identical, there is no copyright infringement. This could happen if, for example, the author of a history of the United States, by accident and without ever reading the work of another author, happens to use the same phrase to describe a key historical figure. Of course, independent creation can be extremely difficult to prove or disprove.

Copyright infringement occurs when someone copies a substantial and material part of another's copyrighted work without permission. The courts have a wide range of discretion in awarding damages—from $200 for innocent infringement up to $100,000 for willful infringement.

Software Copyright Protection

The use of copyright to protect computer software can be tricky because it raises complicated issues of interpretation. For example, a software manufacturer can observe the operation of a competitor's copyrighted program and then create a new program that accomplishes the same result and performs in the same manner. To prove infringement, the copyright holder must show that there is a striking resemblance between the copyrighted software and the new software, a resemblance so close that it could be explained only by copying. However, if the software manufacturer of the new program can establish that it developed the new program on its own without any knowledge of the pre-existing program, there has been no infringement. For example, a software manufacturer could develop a simple program to play tic-tac-toe. Obviously, this game will look a lot like anyone else's tic-tac-toe software, but it could have been developed without copyright infringement. Again, development with no prior knowledge is difficult to prove or disprove.

The Digital Millennium Copyright Act

The Digital Millennium Copyright Act (DMCA) was signed into law in November 1998. The DMCA was written in compliance with the global copyright protection treaty from the World Intellectual Property Organization (WIPO), one of 16 specialized agencies of the United Nations whose goal is to protect the rights of the creators and owners of intellectual property worldwide. The DMCA added new anti-circumvention provisions, making it an offense to engage in an act of circumvention of a technical protection; to develop and provide tools to others that would allow them to access a technologically protected work; and to manufacture, import, provide, or traffic in tools that would enable another to circumvent protection to copy a protected work. These provisions carry both civil and criminal penalties including a maximum of 5 years of imprisonment and/or a fine of up to $500,000 for each offense. Unlike traditional copyright law, the statute does not govern copying, but rather the distribution of tools and software or information that can be used for copyright infringement as well as for legitimate non-infringing uses, such as fair use.

In addition to the Adobe Systems versus ElcomSoft case discussed at the start of the chapter, there have been other cases brought under this act. Several of them deal with the use of software to enable the copying of DVD movies. Read the Legal Overview to get a better understanding of the facts and issues involved in one such case.

Opponents of DMCA say that it gives holders of intellectual property so much power that it actually restricts the free flow of information. For example, under DMCA, Internet Service Providers (ISPs) are required to remove access to Web sites that allegedly break copyright laws—even before the copyright infringement has been proven. Executives from companies that provide Internet access to music and videos are facing legal actions and must gain approval to allow access to content from the music and movie industry or their businesses may fail or be shut down.

 LEGAL OVERVIEW

THE BATTLE OVER CSS

DVDs were introduced in 1996 in the United States as a means of storing motion pictures in digital form. The technology represented a major breakthrough because digital copies made from DVDs do not degrade from generation to generation as videotapes do. In addition, the DVD form was compact and amenable to use on a computer. DVDs have become the format of choice for the distribution of copyrighted movies to consumers, with over 5,000 motion pictures released in that format in the United States. More than 40 new movies are being issued each month, including re-releases of film classics. Some 6 million DVD players have been sold and DVD disc sales exceed one million units per week.

Motion picture companies supported the development of an access control and copy prevention system to prevent the unauthorized reproduction and distribution of motion pictures in the DVD format. The Content Scramble System (CSS) is an encryption-based security and authentication system that requires the use of appropriately configured hardware such as a DVD player or a computer DVD drive to decrypt, unscramble, and play back, but not copy, motion pictures on DVDs. CSS has been licensed to hundreds of DVD player manufacturers and DVD content distributors around the world.

In January 2000, eight movie studios sued *2600: The Hacker Quarterly* for posting the software program DeCSS, capable of breaking the code used to encrypt DVDs on its Web site. United States District Judge Lewis Kaplan granted their request for a preliminary injunction to force the immediate removal of the software program from the quarterly's Web site. At the time, Jack Valenti, president and CEO of the Los Angeles-based Motion Picture Association of America (MPAA represents the interests of the seven largest American movie makers, including Walt Disney, Sony Pictures, MGM, Paramount Pictures, 20th Century Fox, Universal Studio and Warner Brothers) said that "Judge Kaplan's ruling represents a great victory for creative artists, consumers, and copyright owners everywhere. I think this serves as a wake-up call to anyone who contemplates stealing intellectual property."[2] The MPAA subsequently claimed that the illegal copying of DVDs costs the motion picture industry $2.5 billion a year.

In August 2000, Judge Kaplan ruled that *2600: The Hacker Quarterly* is permanently banned from posting or linking to the DeCSS software program.

(continued)

137

The MPAA brought the action under the DMCA, which bans offering or providing technology that may be used to bypass technological means of controlling access to copyrighted works. MPAA argued that DeCSS violated the anti-circumvention provisions of the DMCA. The defendants argued that this interpretation of the law would prohibit the fair use of copyrighted materials and hamper established circumvention of access control measures to reverse engineer and develop interoperable software programs.

In reaching his decision, Judge Kaplan equated DeCSS with a destructive computer virus that had the potential to damage the nation's economy. "In an era where the transmission of computer viruses—which like DeCSS, are simply computer code and thus to some degree expressive—can disable systems upon which the nation depends and in which other computer code is also capable of inflicting other harm, society must be able to regulate the use and dissemination of code in appropriate circumstances," Kaplan said in his decision.[3]

Sources: Ann Harrison, "Judge Bars Posting of DVD Decoding Apps," *Computerworld*, January 24, 2000, accessed at www.computerworld.com; Christine McGeever, "Movie Group Wants Enjoined DVD Site to Stop Hyperlinking," *Computerworld*, April 7, 2000, accessed at www.computerworld.com; Ann Harrison and Todd Weiss, "Judge Blocks Site from Distributing DVD Decryption Software," *Computerworld*, August 18, 2000, accessed at www.computerworld.com.

138

Patents

A **patent** is a grant of a property right to the inventor(s) and is issued by the United States Patent and Trademark Office (USPTO). A patent permits its owner to exclude members of the public from making, using, or selling the claimed invention. It thus enables the inventor to take legal action against those who, without the inventor's permission, manufacture, use, or sell the invention during the period of time the patent is in force. Not only does a patent prevent copying, but, unlike a copyright, it also prevents independent creation. This means that even if someone else invents the same thing independently and with no prior knowledge whatsoever of the patent holder's invention, the second inventor is excluded from use of the patented device without permission of the original patent holder. The rights of the patent extend only to the United States and its territories and possessions, not to foreign countries.

The value of patents to a company cannot be underestimated. For instance, IBM holds about 19,000 United States patents and 34,000 worldwide and estimates that its licensing of patents and technologies generated more than $15 billion in revenue in 2000. Through cross-licensing deals, IBM's patents gain it the rights to use the innovations of other companies while these companies use IBM's technology. Such cross-licensing helps establish IBM's patented technology as a standard in the IT industry, thus encouraging more companies to select IBM products.[4]

To obtain a United States patent, an application must be filed in the USPTO according to strict application requirements. **Prior art** is the existing body of knowledge that is available at a given time to a person of ordinary skill in the art. As part of the patent application process, the USPTO searches the prior art, starting with patents and published material that have already been issued in the same art area. The USPTO will not issue a patent for an invention whose professed improvements are already present in, or are obvious from, the prior art. Although the USPTO employs some 3,000 patent examiners to research the originality of each patent application, it still takes an average of 25 months to process a patent.[5] Such delays can be costly for companies that want to bring patented products to market quickly. Individuals who are trained in the patent process rather than the inventors themselves prepare about 80 percent of all patent applications.

The main body of law governing patents is found in Title 35 of the United States Code. According to sections 103–105 of Title 35, an invention must pass four tests to be eligible for a patent: 1) it must fall into one of the five statutory classes of things that can be patented, 2) it must be useful, 3) it must be novel, and 4) it must be not be obvious to a person having ordinary skill in the art to which said subject matter pertains. The five statutory classes of things that can be patented include processes, machines, manufactures (such as objects made by humans or machines), compositions of matter, and new uses of any of the above. In addition, the United States Supreme Court has ruled that there are three categories of subject matter for which one may not obtain patent protection. These are abstract ideas, laws of nature, and natural phenomena. Mathematical subject matter, standing alone, is also not entitled to patent protection. Thus, Pythagoras could not have obtained a patent for the formula for the length of the hypotenuse of a right triangle ($c^2 = a^2 + b^2$).

Patent infringement occurs when someone makes unauthorized use of another's patent. Unlike copyright infringement, there is no specified limit to the monetary penalty if patent infringement is found. In fact, if the court determines that patent infringement is intentional, up to triple the amount of the damages claimed by the patent holder can be awarded.

A software-related patent claims as all or substantially all of its invention some feature, function, or process embodied in instructions that are executed on a computer. Prior to 1981, the courts regularly turned down requests for such patents, giving the impression that software cannot be patented. In the 1981 *Diamond v. Diehr* case, the Supreme Court denied a patent to Diehr, who had developed the use of a process control computer and sensors to monitor the temperature inside a rubber mold. However, the USPTO interpreted the court's reasoning behind its decision to be that just because an invention utilized software, does not mean that it cannot be patented. Based on this ruling, the courts have slowly broadened the scope of protection available for software-related inventions.

The creation of a new Court of Appeals for the Federal Circuit in 1982 further improved the environment for the use of patents for software-related inventions. This court is charged with hearing all patent appeals and is generally viewed as providing stronger enforcement of patents and more effective punishment, including triple damages for willful infringement.

Since the early 1980s, the USPTO has granted a large number of patents for software-related inventions, with as many as 20,000 software-related patent applications granted each year. Applications software, business software, expert systems, and system software have been patented. In addition, software processes such as compilation processes, edit and control functions, and operating system techniques have been patented. Even electronic font types and icons have been patented.

139

Before obtaining a software patent, the software developer must do a lengthy and expensive patent search. However, even a thorough patent search may not identify all potential infringements because the USPTO's classification system is complex and different software falls under different classifications. If a patent search misses something, there is some risk of an expensive patent infringement lawsuit (the average patent infringement court case cost each side $1.5 million in 1999).[6] The Software Patent Institute is building a database of information documenting the known patented software to assist the USPTO and others in researching prior art in the software arena.

Some software experts think that too many software patents are being granted and that these patents are inhibiting new software development. For example, in September 1999, Amazon.com obtained a patent for "one-click shopping" based on the use of the shopping cart model purchase system for electronic commerce purchasing events. In October 1999, Amazon.com sued Barnes & Noble.com for allegedly infringing on this patent with its "Express Lane" feature. The filing of the suit prompted complaints from critics of "business method" patents, which skeptics deride as overly broad and unoriginal concepts that abuse the patent system and stifle innovation. One-click was considered by some as little more than a simple combination of existing Web technologies. Following preliminary court hearings and the filing of injunctions, Amazon.com and Barnes & Noble settled out of court in March 2002. Examples of the use of the one-click technology were discovered in use before Amazon.com even began business.[7]

Inventors are employing a tactic called defensive publishing as an alternative to filing for patents. Under this approach, a company publishes a description of their innovation in a bulletin, Web site, conference paper, or trade journal. This establishes an idea's legal existence as prior art. Obviously, this provides competitors with access to the innovation. However, because the idea now exists in prior art, competitors will not be able to patent the idea nor charge licensing fees from all other users of the technology or technique. This approach costs mere hundreds of dollars, requires no lawyers, and is fast.

Trade Secret Laws

The Uniform Trade Secrets Act (UTSA) is legislation that was drafted in the 1970s by the National Conference of Commissioners on Uniform State Laws as a model, with the goal of bringing uniformity among the states in the area of trade secret law. The first state to enact the UTSA was Minnesota in January 1981, and 39 additional states plus the District of Columbia have followed Minnesota's lead. In Chapter 2, a trade secret was defined as a piece of information used in a business that represents something of economic value, has required effort or cost to develop, has some degree of uniqueness or novelty, is generally unknown to the public, and that the company has taken strong measures to keep confidential. Similarly, the UTSA defines a trade secret as "information, including a formula, pattern, compilation, program, device, method, technique, or process, that:

- derives independent economic value, actual or potential, from not being generally known to, and not being readily ascertainable by persons who can obtain economic value from its disclosure or use, and

- is the subject of efforts that are reasonable under the circumstances to maintain its secrecy.

Under these terms, computer hardware and software can qualify for trade secret protection by the USTA.

The Economic Espionage Act (EEA) of 1996 imposes penalties as high as $10 million and 15 years in prison for the theft of trade secrets. Before the EEA, there previously was no specific criminal statute to help the FBI pursue the nearly 800 cases of economic espionage by 23 countries under investigation at the time it was enacted.

As with the USTA, information is considered a trade secret under the EEA only if companies take steps to protect it. Indeed, failure to take steps to protect information can weaken any successful EEA prosecution. Trade secret protection begins by identifying all the information that must be protected (ranging from undisclosed patent applications and know-how to market research and business plans) and developing a comprehensive strategy for keeping that information secure.

Actions by a company's employees are the greatest threat to the loss of its trade secrets. Such actions may include accidental disclosure by employees unaware that they're communicating trade secrets to deliberate theft by employees seeking monetary gain. Organizations must educate their employees about the importance of maintaining the secrecy of corporate information. Trade secret information should be labeled clearly as CONFIDENTIAL and be accessible by a limited number of people and on a "need to know" basis. Most organizations have strict policies regarding nondisclosure of corporate information that are documented in employee handbooks and reiterated on a regular basis.

There is a high degree of risk of losing trade secrets when key employees leave the organization. One way companies try to keep their secrets from getting out is by requiring employees to sign nondisclosure clauses in employment contracts. The **nondisclosure clause** requires employees to refrain from revealing company trade secrets. Thus employees cannot take copies of computer programs or reveal the details of software owned by the firm, even when they leave the firm. However, enforcing employment agreements can be difficult.

Another option for keeping secrets safe when employees leave is to have an experienced member of the human resources organization conduct an exit interview with each departing employee. A key step in the interview is to review a checklist dealing with confidentiality issues. Upon conclusion of the exit interview, the employee should sign an "Exit Interview Acknowledgment" statement that reiterates the employee's responsibility not to divulge any trade secrets.

Employers can also use non-compete agreements to protect intellectual property from being used by competitors when key employees leave. Such **non-compete agreements** require employees to not work for any competitors for a period of time, perhaps one to two years. Courts have been asked to settle disputes over non-compete agreements. When doing so, they consider the reasonableness of the restrictions and how those restrictions protect the legitimate interests of the former employer. The courts also consider elements like geographic area and length of time of the restrictions in relation to the pace of the industry. A recent survey by TMP Executive Search revealed that 78 percent of the telecommunications, software, and hardware companies surveyed ask at least some of their employees to sign non-compete agreements when they are hired.[8]

Another approach taken by software manufacturers to protect their products is to use licensing agreements with those who use their software. With such agreements, companies do not actually sell their software to users but license its use instead—the purchaser buys and owns the license, not the software itself. Those who acquire such licenses must agree to do nothing that will reveal the licensing company's trade secrets (such as attempting to reverse engineer the software). They also must agree not to give away or sell copies of the software that has been licensed.

The increasing use of corporate networks and the Internet has heightened the risk of loss of trade secrets stored on computers. A company's information technology organization must take the lead in implementing safeguards (such as firewalls) to limit outside access to corporate computers, encryption for sensitive e-mail, restricted access to servers, and powerful database security to assure that electronic copies of the firm's trade secrets are not accidentally or deliberately released. An area of special concern is the use of computers away from the office by those who travel or take work home on a regular basis. Such computers can be easily stolen and are often used in a careless way. For example, in 1996 it was learned that the CIA director had used his home computer to work on top-secret material at home. The computer was not configured with the special hardware and software needed for classified work and was also used to connect to the Internet, thus creating an opportunity for hackers around the world to steal secret information. After being called before the Senate Armed Services Committee and its probing line of questioning, he received a reprimand and lost his security clearance. He ended up the subject of a very high-profile and damaging investigation and later received a controversial presidential pardon from the departing Bill Clinton.

There are three key advantages that trade secret law has over the use of patents and copyrights in protecting companies from losing control of their intellectual property. First, there are no time limitations on the protection of trade secrets, such as exist with patents and copyrights. Second, there is no need to file any application, make any disclosures, or otherwise disclose a trade secret to outsiders to gain protection—with a patent, after the USPTO issues the patent, competitors can read a detailed description of it. Third, patents can be ruled invalid by the courts, meaning that the then-disclosed invention no longer has patent protection; this risk does not exist for trade secrets. As a result of these advantages, more technology worldwide is protected as a trade secret rather than by patent.[9]

Trade secret law protects only against the *misappropriation* of trade secrets. If competitors come up with the same idea on their own, it is not misappropriation. The law doesn't prevent someone from using the same idea if it was arrived at independently. For example, development of a software program from scratch that duplicates the functionality of another program would be legal; yet, stealing the source code of the same program would be illegal. In addition, trade secret law does not prevent someone from analyzing the end product to figure out the trade secret behind it.

According to the FBI, Fortune 1000 companies suffered losses of more than $45 billion from theft of proprietary information in 1999 alone. One example of the potential for loss of trade secrets occurred in the late 1990s, when two individuals advertised the availability of five stolen prototype Intel CPUs over the Internet. Employees of Cyrix (a competitor of Intel) and Intel worked with the FBI to arrange for the individuals to bring the CPUs to a location in Texas for Cyrix personnel to pretend to inspect and purchase.

One of the CPUs was identified as one of five stolen from Intel in California and the individuals were arrested. Both eventually pled guilty to conspiracy to commit theft of trade secrets. One was sentenced to 77 months imprisonment and a $10,000 fine while the other was sentenced to 60 months imprisonment and a $50,000 fine.[10] In another example, three Chinese citizens were accused of stealing trade secrets from Lucent Technologies in 2002. Two of the three were scientists who worked on the Lucent staff developing the PathStar system for data and voice transmission. They planned to sell a clone of this product to Internet providers in China.[11]

Trade secret protection and patent law differ greatly from country to country. The Philippines provides no legal protection for trade secrets at all. Pharmaceuticals, methods of medical diagnosis and treatment, and information technology items cannot be patented in some European countries. Many Asian countries require transfer of technology held by foreign corporations operating in that country to enterprises controlled by that country. (Coca-Cola closed its doors in India after many years of successful operation to protect the "secret formula" for its soft drink, even though India's population of 550 million represented a huge potential market.) American businesses that seek to operate in foreign jurisdictions or enter international markets must take these differences into account.[12]

Even the hint of legal action against a company for violation of trade secret law can cause extreme financial damage to a company. Shareholder concerns over allegations of industrial espionage reduced the stock market value of Reuters by $1.6 billion during 1998. Subscribers to Reuters rival Bloomberg's system receive a near-constant stream of data about stock, bond, and other financial transactions via special terminals supplied by Bloomberg in New York. With the terminals, users also receive software that provides nearly 6,000 built-in functions for analyzing current and historical data. Reuters' United States subsidiary, Reuters Analytics, was alleged to have improperly induced an outside consultant to obtain proprietary information from Bloomberg, with the result that Bloomberg code ended up in Reuters' software products for displaying stock market data and predicting financial trends. Following six months of FBI investigations and grand jury hearings, the United States Attorney's office announced on July 15, 1999, that it had closed its investigation and concluded that no charges of any sort were warranted. Reuters stock went up that day by $5.50 to $86.

KEY INTELLECTUAL PROPERTY ISSUES

This section will discuss a number of issues relevant to intellectual property and information technology, including reverse engineering, the Uniform Computer Information Transactions Act, competitive intelligence, and cybersquatting.

Reverse Engineering

Reverse engineering is the process of breaking something down in order to understand it, build a copy of it, or improve it. Reverse engineering was originally applied to computer hardware but is now commonly applied to software. Reverse engineering of software involves analysis of an existing software system to create a new representation of the system in a different form or at a higher level of abstraction. To understand what is meant by higher level of abstraction, one must be aware that the classic software development lifecycle is divided into six stages: define requirements, design, code, test, implement, and maintain. If the reverse engineering process begins using the output from the code stage, you can extract design stage information that is at a higher abstraction level in the lifecycle. In other words, design-level details about an information system are more conceptual and less defined than the program code of the same system.

One frequent use of the reverse engineering process for software is converting an application that ran on one vendor's database so that it can run on another's (for example, from dBASE to Access or DB2 to Oracle). Database management systems, such as those mentioned, use their own programming language for application development. As a result, organizations that want to change database vendors are faced with rewriting their existing applications using the new vendor's database programming language. The cost and length of time required for this redevelopment can deter an organization from changing vendors and cause them to miss the benefits that might otherwise result from the conversion to an improved database technology.

Reverse engineering tools can start from the code of the current database programming language and recover the design of (reverse engineer) the information system application. Next, code-generation tools can be used to take the design and produce code (forward engineer) in the new database programming language. This reverse engineering and code generating process greatly reduces the elapsed time and cost to migrate the organization's applications to the new database management system environment. No one challenges the use of the reverse engineering and code-generating process to convert in-house developed applications to a new database management system. After all, the application was developed and is "owned" by the company using it. It is quite another matter, though, to use this process on a purchased software application developed by outside parties. Most information technology managers would consider this action unethical as the software user does not actually own the right to the software. In addition, a number of intellectual property issues would be raised depending on whether the software were licensed, copyrighted, or patented.

A **compiler** is a language translator that converts computer program statements expressed in a source language (such as COBOL, Pascal, or C) into machine language (a series of binary codes of 0s and 1s) that the computer can execute. When a software manufacturer provides a customer with its software, it usually provides it in machine language form. There are also software tools called reverse-engineering compilers or **decompilers** that can read the machine language and produce the source code. For example, REC is a decompiler that reads an executable, machine-language file and produces a C-like representation of the code used to build the program.

The use of decompilers and other reverse engineering techniques can be used to analyze a competitor's program by examining its coding and operation in order to develop a new program that either duplicates the program that has been analyzed or that will interface with that program. Thus, reverse engineering provides a means to gain access to information that another organization may have copyrighted or classified as a trade secret.

The courts have ruled in favor of the use of reverse engineering to enable interoperability. In the early 1990s, video game maker Sega developed a computerized lock so that only Sega video cartridges would work on its entertainment systems.

144

This essentially shut out all competitors from making software for the Sega systems. *Sega Enterprises Ltd. v. Accolade, Inc.* dealt with rival game maker Accolade's use of a decompiler to read the Sega software source code. With the code, Accolade could create new software that circumvented the lock and that was able to run on Sega machines. After an initial lawsuit and appeals, the courts finally ruled that when someone lacks access to the unprotected elements of an original work, and has a "legitimate reason" for gaining access to those elements, disassembly of a copyrighted work is considered to be a fair use under section 107 of the Copyright Act. The unprotected element in this case was the code necessary to enable software to interoperate with the Sega equipment. The court reasoned that to refuse someone the opportunity to create an interoperable product would allow existing manufacturers to monopolize the market, making it impossible for others to compete. This ruling had a major impact on the video game industry, allowing video game makers to create software that would run on multiple machines.

The DMCA explicitly outlaws technologies that can defeat copyright protection devices, but does permit reverse engineering for encryption, interoperability, and computer security research. In spite of the DMCA, court rulings have confused the reverse engineering issue. In addition, some significant lawsuits have been settled out of court, thus eliminating an opportunity to set useful legal precedents. For example, Mattel filed a lawsuit in 2000 against two individuals who reverse engineered its CyberPatrol Web filtering program to create software utilities that revealed the CyberPatrol user's password and displayed a list of blocked Web sites and newsgroups. The utilities essentially gave their users the ability to view and modify the list of blocked sites, thus circumventing the CyberPatrol filtering capabilities. The creators of the utilities then provided copies of them to various Web site operators who posted them for download. Mattel argued that the decompiling of its software and the subsequent posting of the utilities for others to access freely violated copyright law and was prohibited by its software license. A United States District Court judge granted Mattel a temporary restraining order against the posting of the software. Mattel sent notices of the restraining order to the Web sites that had posted the software and requested that they remove the programs and turn over logs of people who had downloaded the banned software. Before the case was resolved, Mattel reached a settlement with the authors of the reverse engineering programs in which the authors agreed to turn over all rights to the reverse engineered software.[13]

Software license agreements increasingly explicitly forbid reverse engineering. Reverse engineering restrictions will also be strengthened by the Uniform Computer Information Transactions Act (see next section), which gives vendors powerful leverage in contract negotiations.

As a result of the increased legislation affecting reverse engineering, some software developers are moving their reverse-engineering projects offshore to avoid United States rules. For example, the developers of Samba software (which helps users access and share UNIX files from Microsoft Windows environments) performed reverse engineering to develop their software because Microsoft doesn't provide documentation of its proprietary protocols. The Samba team avoided reverse engineering issues by having the work done in Europe under the more liberal European Union fair-use laws.[14] Although this is apparently legal, some question the ethics of such action.

The ethics of using reverse engineering can be hotly debated. Some would argue that the use of reverse engineering is fair to enable one company to create software that interoperates with another company's software or hardware and provides a useful function. This is especially true if the original creator of the software refuses to cooperate by providing documentation to help create interoperable software. From the consumer's standpoint, such stifling of competition can increase costs and reduce business options. Reverse engineering can also be a useful tool in detecting bugs and security holes in software. Others argue strongly against the use of reverse engineering. They reason that reverse engineering can uncover the design of software that another has developed at great cost and taken care to protect. To do so unfairly robs the creator of potential future earnings. Without a payoff, what is the business incentive for software development?

Uniform Computer Information Transactions Act (UCITA)

The United States economy has changed greatly since the 1950s, when most of the laws that are the source of current contract law were enacted. These laws focused on the sale of tangible goods; however, intangible products (such as software, databases, and information) and services (software development, Internet access, and hardware and software maintenance) have become an important part of our economy. There currently is no uniform contract law among the states covering intangible products and services. Instead, each state applies its own distinct laws.

In July 1999, the National Commissioners on Uniform State Laws (NCCUSL) voted to approve the Uniform Computer Information Transactions Act (UCITA). NCCUSL is a non-profit association comprised of representatives from each state, the District of Columbia, Puerto Rico, and the U.S. Virgin Islands. It has worked to establish uniformity of state laws since 1892 by drafting and proposing specific statutes in areas of the law where uniformity between the states is desirable and makes it easier to conduct business from state to state. The NCCUSL can only propose statutes—no uniform law is effective until a state legislature adopts it. Thus, the process of establishing uniform rules and procedures is almost guaranteed to take several years at best, even for proposed laws that are broadly accepted and non-controversial.

UCITA is a controversial model act that would apply uniform legislation to software licensing issues. The UCITA must be ratified by each state through its legislature. The various states began in the fall of 1999 to introduce bills that would add the UCITA provisions to their commercial code and the states are at different stages of adoption/rejection of UCITA. A concern is that if only some states adopt UCITA, there still will be no uniform contract law dealing with software licensing issues.

UCITA defines a software license as a contract that grants permission to access or use information subject to conditions set forth in the license. There are three common types of software licenses. **Shrinkwrap licenses** physically accompany the disk or package containing the software program or information. **Click-on** and **active click wrap licenses** are usually transmitted electronically and are activated the instant the user installs the software or accesses the information. The term "shrinkwrap license" is frequently used to encompass all three types of license. Most users agree to the terms of the license without reading them thoroughly in order to continue the software installation process without much delay.

Supporters claim that UCITA will bring a needed uniformity to licensing laws and improve the mass-market distribution of electronic information across the United States. These supporters include the Software and Information Industry Association, information-related companies (such as Microsoft and AOL Time Warner), and a number of large organizations such as Caterpillar, Circuit City, John Hancock Mutual Life Insurance Company, Prudential Insurance, Reynolds Metal, and Walgreens.

Opponents argue that UCITA will increase software and information costs to companies, give consumers less protection against poorly designed software, and increase the power of software vendors. Among the opponents are The American Law Institute (a collection of law scholars and practitioners), the 4CITE (a broad-based coalition of end users and developers of computer and information technology), The Association for Computing Machinery, the Computer and Communications Industry Association, Computer Professionals for Social Responsibility, the Digital Future Coalition, the Electronic Frontier Foundation, the International Communications Association, and the Society for Information Management. Specific concerns include the following:

- UCITA validates shrinkwrap or click-on licenses that give the purchaser no room to negotiate because the terms of such a contract are not available for review until after purchase.
- UCITA allows contracts to prohibit transfer of software from one purchaser to another, even in the case of mergers or acquisitions among purchasers of a given software package.
- UCITA allows for remote disabling of software on the purchaser's own computer in the case of a disagreement, giving the producer too much control before a disagreement has been resolved in a legal forum.
- UCITA defines information so broadly that it could mean that material once readily available through the government documents or public domain may soon become restricted simply because it is offered as a computer file.

The value of the UCITA is still very much under debate. The issues discussed in this chapter are ones that each state legislature must consider when deciding whether to adopt the act. It's supported by vendors but opposed by a diverse coalition of end users, libraries, consumer groups, and over two-dozen state attorneys general, who believe the law gives vendors too much power on contracts. They want the law scrapped. Only two states, Virginia and Maryland, have adopted UCITA as of June 2002. As a result, the group that drafted the UCITA is working on a new series of proposals to change the complex law in an attempt to increase its acceptance.

Competitive Intelligence

Competitive intelligence is the gathering of legally obtainable information that will help a company gain an advantage over its rivals. For example, some companies have individuals monitor the public announcements of property transfers to detect any plant or store expansions of a competitor. An effective competitive intelligence operation requires the continual gathering, analysis, and evaluation of data with controlled dissemination of *useful* information to decision makers. Competitive intelligence is often integrated into a company's strategic plans and decision making. Many companies such as Eastman Kodak, Monsanto, and United Technologies have established formal competitive intelligence departments. Some companies have even employed former Central Intelligence Agency (CIA) analysts to conduct competitive intelligence.

Competitive intelligence is not **industrial espionage**, which employs illegal means to obtain business information not available to the general public. In the United States, industrial espionage is a serious crime that carries heavy penalties. Almost all the data needed for competitive intelligence can be collected from careful examination of published information sources or interviews, such as those shown in Table 6-1. By coupling this data with analytical tools and expertise in the industry, an experienced competitive intelligence analyst can deduce many significant facts.

Table 6-1   Common sources of competitive intelligence

10-K or annual report

SC 13D acquisition

10-Q or quarterly reports

Press releases

Promotional materials

Web site (many of the items in this list can be found here)

Analyses performed by the investment community (such as a Standard and Poor's stock report)

A Dun & Bradstreet credit report

Interviews with suppliers, customers, and former employees

Calls to the competitor's customer service group

Articles in the trade press

Environmental impact statements and other filings associated with the expansion or construction of a new plant

Patents

Competitive intelligence gathering has become enough of a science and an important enough skill that a number of academic programs have been developed for it. The Fuld-Gilad-Herring Academy of Competitive Intelligence and Boston's Simmons College each offer a competitive intelligence professional (CIP) certification program. Drexel University offers an online competitive intelligence program. Several colleges and universities also offer competitive intelligence courses or entire programs, including Brigham Young University, Mercyhurst College, Mount St. Mary's College, and Wright State University.[15]

Without proper management safeguards, the process of gathering competitive intelligence can cross over into the "gray area" between competitive intelligence and industrial espionage and can involve the use of "dirty tricks." One frequently employed dirty trick is to go into a bar near a competitor's plant or headquarters, strike up a conversation, and ply people for information after their inhibitions have been weakened by alcohol. Most people would consider such tactics unethical; others would say they are acceptable tactics to use when gathering competitive intelligence.

Competitive intelligence analysts must use care and avoid doing anything that is blatantly unethical or illegal, including lying, misrepresenting, stealing, bribing, or eavesdropping with illegal devices. See Table 6-2, for useful guidelines. The preferred answer to each question in the checklist is "Yes." Failure to act prudently can get analysts and their companies into serious trouble, as the two examples discussed after the table show.

Table 6-2	A manager's checklist for running an ethical competitive intelligence operation

Questions	Yes	No
Has your competitive intelligence organization developed a mission statement, objectives, goals, and a code of ethics?	____	____
Has your company's legal department approved the mission statement, objectives, goals, and code of ethics?	____	____
Do analysts understand the need to agree to and abide by the competitive intelligence organization's code of ethics, corporate policies, and corporate code of ethics?	____	____
Is there a rigorous training and certification process for analysts?	____	____
Do analysts understand all applicable laws, domestic and international, including the Uniform Trade Secrets Act and the Economic Espionage Act, and do they understand the critical importance of abiding by them?	____	____
Do analysts disclose their true identity and organization prior to any interviews?	____	____
Do analysts understand that everything your firm is to learn about the competition must be obtained by legally available means?	____	____
Do analysts respect all requests for anonymity and confidentiality of information?	____	____
Has your company's legal department approved the processes used by analysts to gather data?	____	____
Do analysts provide honest recommendations and conclusions?	____	____
Is the use of third parties to perform competitive intelligence activities carefully reviewed and managed?	____	____

The Procter & Gamble Company (P&G) admitted publicly in August 2001 that it gained information on competitor Unilever's hair-care business in a way that, while not illegal, was not in keeping with P&G principles and policies. It seems that competitive intelligence managers at P&G had hired a contractor, who in turn hired several subcontractors to spy on its competitors in the hair-care business. Unilever was the primary target. The hair-care business is worth billions of dollars annually to both companies. Unilever markets brands such as Salon Selectives, Finesse, and Thermasilk, while P&G manufactures Pantene, Head & Shoulders, and Pert.

In at least one instance, the espionage included going through dumpsters on public property outside Unilever corporate offices in Chicago. In addition, it is alleged that competitive intelligence operatives also misrepresented themselves to Unilever employees, suggesting that they were market analysts. (P&G confirms the dumpster diving, but it denies that misrepresentation took place.) The operatives captured critical information about Unilever's brands, including new-product rollouts, selling prices, and operating margins.

When senior P&G officials discovered that a firm hired by the company was operating unethically, P&G immediately stopped the campaign and fired the three managers responsible for hiring the firm. P&G then did something unusual—it blew the whistle on itself and confessed to Unilever, the primary target of the operation. P&G returned the stolen documents to Unilever and started negotiations with Unilever to set things straight. The P&G chairman of the board was personally involved in ensuring that none of the information obtained was ever used in any P&G business plans.[16] Several weeks of high-level negotiations between P&G and Unilever executives led to a secret agreement between the two companies. It is believed that P&G paid tens of millions of dollars to Unilever. In addition, several hair-care product executives were transferred to other units within P&G.[17]

Experts in competitive intelligence agree that the firm hired by P&G crossed the line of ethical business practices when they sorted through Unilever's garbage. However, they also give P&G credit for going to Unilever quickly after it discovered the damage. Such prompt action was seen as the best way to put things to rest. If no settlement had been reached, of course, Unilever could have taken P&G to court, where embarrassing details would have been revealed, causing further negative publicity for a company that is generally perceived as a highly ethical one. Unilever also stood to lose from a public trial. The trade secrets at the heart of a case often get revealed in depositions and other documents demanded during the trial, which devalues the proprietary data.

In another instance of an overzealous attempt to gain competitive intelligence, Time Warner Cable (now AOL Time Warner) admitted that it offered to reward its Houston employees for placing fake orders for high-speed Internet access. The goal was to find out in which geographical regions Southwestern Bell, a rival Internet Service Provider (ISP), was able to offer service and then exploit any weaknesses uncovered. Time Warner Cable now concedes that the practice violated its own corporation's ethical standards. The tactic used amounted to paying employees to gather information that was not publicly available and doing it in such a manner as to create a lot of extra effort and expense for its competitor. Southwestern Bell general counsel filed formal complaints with the Federal Communications Commission (FCC) and the Texas Public Utility Commission.[18]

Cybersquatting

A **trademark** is anything that enables a consumer to differentiate one company's products from another's. The trademark may be a logo, package design, phrase, sound, or word. Consumers are often unable to examine goods (or services) to determine their quality or source. Instead, they must rely on the labels attached to the products offered for sale. Trademark law gives the producer who regularly uses a mark the right to prevent others from using the same mark or a confusingly similar mark.

There is a federal system that stores trademark information that merchants can consult to avoid adopting marks that have already been taken. Merchants seeking trademark protection apply to the USPTO if they are currently using the mark in interstate commerce or if they can demonstrate a true intent to use the mark in commerce. Trademark protection lasts for as long as a mark is in use.

Companies that want to establish an Internet presence know that the best way to capitalize on the strength of their brand names is to make their names part of the domain names for their Web sites. When Web sites were first being established, there was no procedure for validating the legitimacy of the request for a Web site name and names were given out on a first-come, first-served basis. **Cybersquatters** registered domain names for famous trademarks or company names to which they had absolutely no connection, with the hope that the company owning the trademark would buy the domain name for a large sum of money.

The main tactic used by organizations to circumvent cybersquatting is to protect a trademark by registering, as soon as the organization knows it wants to develop a Web presence, numerous domain names that reflect its possible variations (that is, UVXYZ.com, UVXYZ.org, and UVXYZ.info). In addition, trademark owners to whom non-English speaking customers are important should register their names in multilingual form as well. Registering additional domain names is far less expensive than attempting to force cybersquatters to change or abandon their domain names.[19]

The Internet Corporation for Assigned Names and Numbers (ICANN) is a nonprofit corporation responsible for the management of the domain name system. ICANN is in the process of adding seven new top-level domains (.aero, .biz, .coop, .info, .museum, .name, and .pro) to the Internet's domain-name system. Current trademark holders will be given time to assert rights to their trademarks in the new top-level domains before registrations are opened to the general public. ICANN also has a Uniform Domain Name Dispute Resolution Policy under which most types of trademark-based domain name disputes must be resolved by agreement, court action, or arbitration before a registrar will cancel, suspend, or transfer a domain name. The ICANN policy is designed to provide for the fast, relatively inexpensive arbitration of a complaint by a trademark owner that a domain name was registered or used in bad faith.

The Anti-cybersquatting Consumer Protection Act (ACPA), enacted in 1999, allows trademark owners to challenge foreign cybersquatters who might otherwise be beyond the jurisdiction of United States courts. Under this act, trademark holders can seek civil damages of up to $100,000 from cybersquatters that register their trade names or similar-sounding names as domain names. It also makes it easier for trademark owners to challenge the registration of their trademark as a domain name even if the trademark owner has not created an actual Web site.[20] For instance, a federal court judge in Boston, Massachusetts, ordered a Canadian cybersquatter to vacate the Internet address of a Cambridge, Massachusetts, search engine, NorthernLight.com. It was one of hundreds of Web sites he occupied. The cybersquatter had claimed that the United States court had no jurisdiction in Canada. The judge ruled that the Canadian site did, in fact, infringe upon the firm's trademark, that the cybersquatter had violated the ACPA, and that this act extended the jurisdiction of the United States court to rule in the case of a non-United States citizen infringing on the rights of a United States company.[21]

Summary

1. What does the term "intellectual property" encompass and why are companies so concerned about protecting it?

    Intellectual property is a term used to describe works of the mind, such as art, books, films, formulae, inventions, music, and processes that are distinct and that are "owned" or created by a single entity. Copyrights, patents, trademarks, and trade secrets provide a complex body of law regarding the ownership of intellectual property. Intellectual property represents a large and valuable asset to most companies. If not protected, it is possible for other companies to copy or steal these assets, resulting in significant loss of revenue and loss of competitive advantage.

2. What are the strengths and limitations of using copyrights, patents, and trade secret laws to protect intellectual property?

    A copyright grants the author of an original work the exclusive right to distribute, display, perform, or reproduce the work in copies, prepare derivative works based upon the work, and grant the exclusive right to do any of these actions to others. Copyright law has proven to be extremely flexible in covering new technologies: software, videogames, multimedia works, and Web pages can all be protected by copyright. However, evaluation of the originality of a work can be problematic and has given rise to much litigation.

    A patent enables the inventor to take legal action against those who, without the inventor's permission, manufacture, use, or sell the invention during the period of time the patent is in force. Not only does a patent prevent copying, but it also prevents independent creation (an allowable defense to a copyright infringement claim). Unlike copyright infringement, for which monetary penalties are limited, if the court determines that patent infringement exists and is intentional, up to triple the amount of the damages claimed by the patent holder can be awarded.

    To qualify as a trade secret, a piece of information must have economic value and must not be readily ascertainable. In addition, the trade secret owner must have taken steps to maintain its secrecy. Trade secret law doesn't prevent someone from using the same idea if it was arrived at independently or from analyzing an end product to figure out the trade secret behind it. However, there are three key advantages that trade secret law has over the use of patents and copyrights in protecting companies from losing control of their intellectual property: there are no time limitations on the protection of trade secrets, such as exist with patents and copyrights; there is no need to file any application or otherwise disclose a trade secret to outsiders to gain protection; and there is no risk that a trade secret might be found to be invalid in court.

3. What is reverse engineering and what are the issues associated with applying this technique to uncover the trade secrets of a competitor's software program?

    Reverse engineering is the process of breaking something down in order to understand it, build a copy of it, or improve it. Reverse engineering was originally applied to computer hardware but is now commonly applied to software. In some situations, it can be considered unethical because it provides a means to gain access to information that another organization may have copyrighted or classified as a trade secret. Recent court rulings, shrinkwrap license agreements for software that forbid reverse engineering, and reverse engineering restrictions in the UCITA, have made reverse engineering a riskier proposition in the United States.

4. What is the purpose of the UCITA and what is its potential future impact on software licensing?

    UCITA is a controversial model act that would apply uniform legislation to software licensing issues. The UCITA defines a software license as a contract that grants permission to access or use information subject to conditions set forth in the license. Supporters claim that UCITA will bring a needed uniformity to licensing laws and improve the mass-market distribution of electronic information across the United States. Opponents argue that UCITA will increase software and information costs to companies, give consumers less protection against poorly designed software, and increase the power of software vendors.

5. What is the essential difference between competitive intelligence and industrial espionage, and what are several sources of competitive intelligence information?

    Competitive intelligence is not industrial espionage, which employs illegal means to obtain business information not readily available to the general public. In the United States, industrial espionage is a serious crime that carries heavy penalties. Almost all the

data needed for competitive intelligence can be collected from careful examination of published information sources or interviews. Competitive intelligence analysts must use care and avoid doing anything that is blatantly unethical or illegal, including lying, misrepresenting, stealing, bribing, or eavesdropping with illegal devices.

6. What strategy should be used to protect an organization from cybersquatting?

    The main tactic used by organizations to circumvent cybersquatting is to protect a trademark by registering, as soon as the organization knows it wants to develop a Web presence, numerous domain names that reflect its possible variations. In addition, trademark owners to whom non-English speaking customers are important should register their names in multilingual form as well.

Review Questions

1. What types of works can be copyrighted or patented?

2. What is the fair use doctrine? What are the four factors courts consider in determining whether a particular use of copyrighted property is a fair use and can be allowed without penalty?

3. What is meant by independent creation? Is it an allowable defense to a copyright infringement claim? To a patent infringement claim? To a trade secret infringement claim?

4. Define trade secret. What actions must an organization take to ensure its ability to file a successful Economic Espionage Act claim?

5. What options do companies have to protect their trade secrets?

6. What is a trademark? What is its purpose?

7. What is a cybersquatter?

8. What is reverse engineering as it applies to hardware or software?

9. Identify two events that led to the increase of the number of software patents starting in the early 1980s.

10. What is the difference between competitive intelligence and industrial espionage?

11. Under what laws might someone be punished for engaging in industrial espionage?

12. What is the Digital Millennium Copyright Act (DMCA)? What are some of the specific concerns about the DCMA?

13. What is prior art and what role does it play in the patent process?

Discussion Questions

1. Some people believe that the use of copyrights and patents reduces the competition among software manufacturers and is bad for the industry. Give three reasons why this may be true. Next, take the opposite side of the argument and give three reasons why the use of copyrights and patents are good for the software industry.
2. Check the status of adoption of UCITA in your state in particular and across the United States in general and write a paragraph summarizing your findings.
3. What are the advantages for the software purchaser of having uniform legislation to deal with software licensing issues? What are the advantages for the software manufacturer?
4. What is the best way for a software manufacturer to protect new software? Why?
5. What actions should a company take to avoid problems from cybersquatters?
6. How might a corporation use reverse engineering to convert to a new database management system? How might a corporation use reverse engineering to uncover the trade secrets behind a competitor's software?
7. What recent developments have challenged the use of reverse engineering of software in the United States? Is this development good or bad for the software industry? Why?
8. Compare and contrast the key issues in the Sega and Mattel reverse engineering lawsuits.

What Would You Do?

1. You are a senior vice-president for marketing at a large, highly successful software manufacturer. You have been asked by a national software manufacturer's organization to add your company's name to a list of supporters of UCITA. Would you agree to do so? What might be the repercussions to your firm, both good and bad, from supporting this act?

2. Because of the dollar amount, the approval of the vice-president of finance at your company was required for a $500,000 purchase order for hardware and software to upgrade the servers used to store data for the product development department. Everyone in product development had expected an automatic approval and they were shocked and disappointed when the purchase order request was turned down. The reason given was that the business benefits of the expenditure were not clear. Realizing that she will now need to develop a more solid business case for the order, the vice-president of product development has come to you for help. As the manager of corporate security, what questions would you ask her about this proposed expenditure? Can you identify any arguments related to the protection of intellectual property that might help strengthen the business case for this expenditure?

3. You have been asked to head up your company's new competitive intelligence organization. What actions would you take to ensure that its members violate neither the applicable laws nor the company's own ethical guidelines and policies?

4. You are the manager for software development of a major, multinational software manufacturing company. You have been asked to prepare a proposal to form a reverse engineering group of over 50 software engineers who will use the latest tools and methodologies to study competitors' software in order to identify future enhancements for your own company's products. Would you support such an effort? If not, why not? If so, what restrictions might you place on this group?

Cases

1. Napster Battles Recording Industry

Napster is a Web-based service founded in 1999 by Shawn Fanning, then an 18-year old freshman at Boston's Northeastern University. His goal was to provide music enthusiasts with an easy-to-use, high-quality service for discovering new music and communicating their interests to other members of the Napster community. The service essentially gave those who downloaded its software access to the music on the computers of all the other Napster users who were willing to share their music.

A first-time Napster user would access the Napster Web site to register and create a username and password. The user could then download Napster software, including a search engine to determine what files were available on other users' hard drives and software to exchange music files over the Internet. To make their own music available to others, Napster users would transfer their own CDs into MP3 computer files and store them on their own hard drive using software that is widely available.

The Napster operation relied on a central, constantly updated index of all music available to Napster users. This index was stored on a Napster server that acted like a traffic cop, directing a user's request for songs to other users' hard drives. Users simply typed in a song title or the name of an artist, and Napster generated a list of other users who already had it. The user would then click a selection to copy the file from another user's hard drive to their own.

There were some problems with the process. Downloading a song could take several minutes, even for users with 56-kbps modems or cable connections to the Internet. It might also take several attempts to connect to another user. The content provided by other Napster users was not always reliable—the audio quality could be poor, songs were frequently mislabeled, and incomplete versions were common.

Despite the problems, Napster was wildly popular and became one of the fastest growing Internet Web services ever—growing to over 50 million users within its first 15 months of existence. At its peak, more than 9 million people per month used its services.

The recording industry strongly objected to Napster because it enabled users to copy music from each other free over the Internet. As a result, the Recording Industry Association of America (RIAA) and its member record labels sued Napster in December 1999. They requested a preliminary injunction against Napster, forcing it to prevent users from exchanging copyrighted music. In July 2000, Judge Marilyn Patel of the Federal District Court in Northern California granted the injunction. However, just two days later, a two-judge panel from the Ninth Circuit Court issued an emergency stay to the injunction.

Over the next few months, a number of outside interests, including the Consumer Electronics Association and the Association of American Physicians and Surgeons, filed court briefs on behalf of Napster's position. Although these organizations did not necessarily support Napster's business, they requested that the court not limit the exchange of copyrighted works and information over the Internet too narrowly.

In October 2000, a three-judge United States Court of Appeals Court for the Ninth District heard the Napster case. In February 2001, this court ruled that Napster encourages and helps in the wholesale infringement of copyrights. Their decision largely affirmed the initial lower court ruling.

The United States Court of Appeals judges ruled that not only does the evidence show that Napster encourages copyright infringement, but also that individual Napster users are infringing copyrights. Users made unauthorized copies to save the expense of purchasing additional copies, and therefore, the exchange of copyrighted music cut into the record industry's sales. The court clearly rejected Napster's argument that individuals who exchange music on Napster for their personal use are exercising a fair use of the content.

The court also did not agree with Napster's argument that it was protected by a 1984 Supreme Court decision that found Sony not liable for cases where consumers used its video recorders to copy television shows. In that case, the Supreme Court found that most VCR owners used their machines to tape a copyrighted television show for later personal viewing. The court said this practice of "time-shifting" was a fair use. The Sony case differs from Napster in that the vast majority of VCR users did not distribute to the general public television broadcasts they taped. After a Napster user listed a copy of the music he already owned on the Napster system in order to access the music from other users, the song became available to millions of other individuals, not just the original music owner.

The appeals court did hold out the prospect that the Napster system could be adapted in ways that would make it legally viable as an online music distributor. To that end, the appeals court judges requested that the lower court fashion an injunction that required Napster to stop the exchange of copyrighted works, but that respected Napster's technological limits to police activity on its site. The court also ruled that the record companies must provide notice to Napster of which works are being infringed. The new injunction must devise a remedy that takes into account the technological limits of the service because Napster cannot read files stored on the computers of individual users.

In March 2001, Napster said it would begin to prevent users from exchanging some copyrighted songs online. The file names that Napster would begin to filter out of its system are those reported to Napster by aggrieved copyright holders. However, Napster made it clear that a given song might be known by specific names that Napster users themselves give to songs. A misspelling, intentional or unintentional, could fool the blocking software. For example, a user looking for Bob Seger's song "Night Moves" would be blocked from getting the song, while one typing in "Nite Movz" could turn up numerous entries.

Although Napster was struggling with its legal difficulties, the five major record companies were scrambling to create their own profitable, legal,

online music subscription services. Bertelsmann A.G., the EMI Group, and AOL Time Warner struck a deal with RealNetworks, an Internet audio and video distribution company, to make music available online through a legal music subscription service called MusicNet. Sony and Vivendi Universal Music Group announced that their Internet subscription service Pressplay would provide access to almost 100,000 songs and provide additional information such as lyrics and artist biographies. Vivendi Universal also paid $372 million for MP3.com, the pioneering music Web site that Vivendi's own music unit had previously sued for copyright infringement.

The efforts of the five major record companies to join forces and sell music over the Internet has not gone unnoticed by federal regulators and legislators who want to ensure that the companies do not engage in anticompetitive behavior. The Justice Department has begun preliminary investigations into MusicNet and Pressplay. Legislators are considering legislation that would require companies who license music for sale to one Internet site to make the music available to other sites under the same terms. Some smaller sites and industry observers have expressed concern that the companies would limit how they licensed music online to control the market and ensure huge profits.

In June 2001, Napster was working to close an agreement to license music from MusicNet in a pay version of its service. However, in July 2001, Judge Marilyn Hall Patel ruled that Napster could not resume allowing the exchange of audio files until the system was at a point where it had been tested and shown to have the ability to keep out unauthorized copyrighted works.

In September 2001, Napster signed a preliminary agreement with the National Music Publishers' Association (NMPA) and the songwriters the group represents to end some of the litigation pending against it. Under the agreement, Napster will pay $26 million in damages for past, unauthorized use of music and will advance another $10 million for future licensing royalties from its yet-to-be-launched, fee-based service. Songwriters will also be entitled to a portion of Napster's revenue generated by song sales on its Web site.

As Napster struggles to convert to a new pay-for-service business model, file-sharing programs such as BearShare, Freenet, Gnutella, LimeWire, MusicCity, Napigator, and OpenGap are gaining in popularity. By the same token, the new Napster service is being met with a great deal of skepticism. With new fees and restrictions, Napster will be less appealing to potential users than in the past.

As late as May 2002, Napster was still involved in the original lawsuit with the RIAA that had resulted in Napster being shut down. It also had had no success in making deals with artists to offer their music online. In June 2002, Napster filed for bankruptcy and subsequently Bertelsmann AG acquired the company for $8 million.

Questions:

1. What are the key differences between the original Napster business model and MusicNet's approach?

2. What is your opinion on the ethics of using Napster before it was forced to change its business model?

3. Do you agree that the use of the Napster service differs significantly from those who use a VCR to record copyrighted TV programs? Why or why not?

Sources: adapted from Ashlee Vance, "Napster Serenades Songwriters, Ready to End Lawsuit," IDG News Service, September 25, 2001; Industry Standard Staff, "Judge Orders Napster to Stay Down," *The Industry Standard*, July 12, 2001, accessed at www.industrystandard.com; John Markoff, "Record Companies Seek Fees for Net Music," *The New York Times on the Web*, April 3, 2001, accessed at www.nytimes.com; Kenneth Li and Hane C. Lee, "Vivendi Universal Buying MP3.com for $372 Million," *The Industry Standard*, May 21, 2001, accessed at www.industry-standard.com; Amy Harmon, "Congress Getting a Preview of Online Music Service," *The New York Times on the Web*, May 13, 2001, accessed at www.nytimes.com; Matt Richtel, "Plans to Sell Music on the Internet Raise Antitrust Concerns," *The New York Times on the Web*, August 7, 2001,

accessed at www.nytimes.com; Scarlett Pruitt, "Bertelsmann Buys Napster for $8 Million," *Computerworld*, May 20, 2002, accessed at www.computerworld.com; Associated Press, "Napster Seeks Bankruptcy Protection," June 3, 2002, accessed at www.usatoday.com.

2. Lotus v. Borland

By the early 1990s, conflicting court decisions over the previous decade had confused software developers in regard to the issue of the "look and feel" of software. Must every software product have its own unique look and feel? Could a company improve its products by incorporating new features that existed in competing products? Was it legal to build and market a clone—a product externally identical to another product? The *Lotus v. Borland* lawsuit and associated appeals lasted over five years—going all the way to the Supreme Court. They set a precedent that clarified the limits of software copyright protection.

The first electronic spreadsheet on the market had been Daniel Bricklin's VisiCalc, introduced in 1979. This program enabled people to easily prepare budgets, forecast profits, analyze investments, and summarize tax data. VisiCalc was a huge success, selling more than 100,000 copies its first year on the market, this at a time when there were only a few million PC owners. For the first time, users found an application compelling enough to make them want to buy a computer. As a result, VisiCalc and other early spreadsheet programs are credited with being a key catalyst for the PC revolution. Neither the VisiCalc software nor the concept of a spreadsheet program were patented or copyrighted.

Mitch Kapor and Jonathan Sachs founded Lotus Development Corporation in 1982. In 1983, Lotus released its flagship spreadsheet product, Lotus 1-2-3, which, like VisiCalc, enabled users to perform accounting functions electronically on a computer. Lotus 1-2-3 was highly successful and generated sales of $53 million in its first year and $156 million the next year.

Users manipulated and controlled the Lotus 1-2-3 program by using a series of menu commands, such as "Copy," "File", "Print," and "Quit." The commands could be selected by either highlighting them on the screen or by typing their first letter. Lotus 1-2-3 also allowed users to write a macro, a user-designated series of command choices activated by a single macro keystroke. This caused the program to recall and perform the designated series of commands automatically.

Borland was founded in 1983 with the objective of developing a program superior to Lotus 1-2-3. After several years of development, Borland released its Quattro spreadsheet program to the public in 1987. The Borland software gave users a choice of how they could communicate with Borland's spreadsheet program—by using menu commands designed by Borland, or by using the commands and command structure found in the popular Lotus 1-2-3 program. This capability would make it easier to attract Lotus 1-2-3 users to the Borland software. To provide this dual functionality, Borland copied only the words and structure of Lotus's menu command hierarchy, but not any of its competitor's computer code.

Lotus sued Borland in 1990, charging that the execution of commands in Borland's spreadsheet program copied the look and feel of the Lotus 1-2-3 interface. In August 1992, the United States District Court found that Borland included in its spreadsheet "a virtually identical copy of the entire Lotus 1-2-3 menu tree" and infringed the copyright of Lotus. The judge held that the menu command hierarchy—the selection of menu items and their particular arrangement in an inverted tree structure—was a protectable element of the program that Borland had infringed. In addition, the judge ruled that Borland's "key reader"—a program feature that enabled the Borland spreadsheet program to interpret and execute Lotus 1-2-3 macros—was infringing because it employed a table that reproduced the entire Lotus 1-2-3 menu command hierarchy, with the first letter of each command substituted for the full command name.

The court subsequently entered an injunction against Borland against further sales or distribution of then current versions of Borland's spreadsheet products. In response, Borland shipped new versions of its spreadsheet that did not include the features found to be infringing. Borland also appealed the court's decision. The district court reaffirmed its decision in July 1993.

Deciding to get out of the spreadsheet business, Borland sold its Quattro spreadsheet software to Novell Inc. in March 1994. One year later, the United States Court of Appeals for the First Circuit reversed the district court ruling that Borland had infringed the copyright of Lotus 1-2-3. The Court of Appeals found that the menu command hierarchy was a "method of operation" that is excluded from copyright protection under U.S.C. Title 17 section 102(b), because "the Lotus menu command hierarchy provides the means by which users control and operate Lotus 1-2-3." The ruling was welcomed by Borland, which would have found it extremely difficult to pay the estimated $100 million in damages sought by Lotus.

The case was appealed to the Supreme Court and in January 1996, five years after the start of the initial copyright infringement suit with Lotus, the Supreme Court made its ruling. The court affirmed the decision by the First Circuit Court of Appeals, which had ruled in Borland's favor.

The *Lotus v. Borland* case was significant for the software industry, which had been riddled with infringement lawsuits due to the ambiguities in copyright law. The ruling made it clear that software copyrights could be successfully challenged, thus further discouraging the use of copyright as a means to protect software innovations. As a result, software developers were forced to go through the much more difficult and expensive patent process to protect their software products. These additional costs would either be absorbed by developers or passed on to users.

Borland is still in business today and provides products and services targeted to software developers, including a Borland version of the C++ programming language and the database software packages Visual dBASE, Paradox, and InterBase. Recent annual sales were in the neighborhood of

$200 million. IBM purchased Lotus in 1995 for $3.5 billion in cash. Lotus 1-2-3 is still widely used, although it is not as popular as the Microsoft Excel spreadsheet program.

Questions:

1. Go to your school's computer lab or a PC software store and experiment with current versions of any two of the Quattro, Excel, and Lotus 1-2-3 spreadsheet programs. Write a brief paragraph summarizing the similarities and differences in the "look and feel" of these two programs.

2. It took the justice system several years to reverse its initial decision and rule in favor of Borland. What impact on the software industry did this delay have? How might things have been different if Borland had received an initial favorable ruling?

3. Assume that you are the manager in charge of Borland's software development and, assuming perfect 20-20 hindsight, what different decisions would you have made in regard to Quattro?

Sources: adapted from William Brandel, "'Look and Feel' Reversal Re-ignites Copyright Fight," *Computerworld*, March 13, 1995; Martin A. Goetz, "Copycats or Criminals?", *Computerworld*, June 12, 1995; "Borland Prevails in Lotus Copyright Suit," January 16, 1996, Borland Press Release, accessed at www.borland.com.

160

Endnotes

[1] The Associated Press, "Other Music Swapping Services," *The New York Times on the Web*, February 13, 2001, accessed at www.nytimes.com.

[2] Ann Harrison, "Judge Bars Posting of DVD Decoding Apps," *Computerworld*, January 24, 2000, accessed at www.computerworld.com.

[3] Ann Harrison and Todd Weiss, "Judge Blocks Site from Distributing DVD Decryption Software," *Computerworld*, August 18, 2000, accessed at www.computerworld.com.

[4] Don Steinberg, "The Patent King", *Smart Business*, April 2001, pg. 75.

[5] Carol Pickering, "Patently Absurd," *Business 2.0*, May 29, 2001, pg. 28–30.

[6] Carol Pickering, "Patently Absurd," *Business 2.0*, May 29, 2001, pg. 28–30.

[7] Beth Cox, "One Settlement Over One Click," *Internet News*, March 7, 2002, accessed at www.atnewyork.com.

[8] Elisabeth Goodridge, "Keeping Ideas In-House," *Information Week*, July 2, 2001, pg. 59.

[9] Karen A. Magri, "International Aspects of Trade Secret Law," accessed at www.carolinapatents.com on October 18, 2001.

[10] United States of America v. Steven Hallsted and Brian Pringle, Criminal Case No. 4: 98M37 (E.D. Texas 2/26/98).

[11] Associated Press, "Trade-Secret Case Is Expanded," *The New York Times on the Web*, April 12, 2002, accessed at www.nytimes.com.

[12] Karen A. Magri, "International Aspects of Trade Secret Law," accessed at www.carolinapatents.com on October 18, 2001.

[13] Ann Harrison, "Cyber Patrol Case Tests Reverse-Engineered Apps," *Computerworld*, April 3, 2000, pg. 6.

[14] Ann Harrison, "Battle Brews Over Reverse Engineering," *Computerworld*, May 4, 2000, accessed at www.computerworld.com.

[15] Matthew Boyle, "The Prying Game," *Fortune*, September 17, 2001, accessed at www.fortune.com.

16 Procter & Gamble Press Release, September 6, 2001, accessed at www.pg.com.

17 Andy Sewrer, "P&G's Covert Operation," *Fortune*, September 17, 2001, accessed at www.fortune.com.

18 "Keeping Your Nose Clean While Sniffing Competitive Info," *Business Week Online*, June 2, 2000, accessed at www.businessweek.com.

19 Rose Auslandeer and Douglas Royce, "New Domain Name Concerns for Trademark Owners," *Computerworld*, October 19, 2000, accessed at www.computerworld.com.

20 Rose Auslandeer and Douglas Royce, "New Domain Name Concerns for Trademark Owners," *Computerworld*, October 19, 2000, accessed at www.computerworld.com.

21 Keith Regan, "Infamous Cybersquatter Shut Down by Judge," *E-Commerce Times*, April 14, 2000, accessed at www.ecommercetimes.com

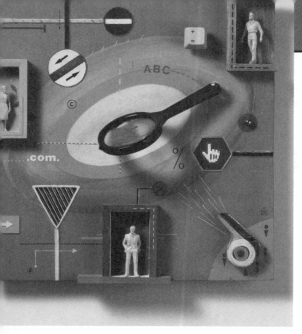

Chapter 7

SOFTWARE
Development

"[Customers should be able to count on] computing that is as available, reliable, and secure as electricity, water services, and telephony."[1]
— Bill Gates in a 2002 e-mail to Microsoft employees

In January 2002, Bill Gates sent an e-mail to Microsoft employees saying that there must be a company-wide emphasis on developing high-quality code even if it comes at the expense of adding new features. The following month, 8,000 Microsoft employees in the Windows division were ordered to drop what they were doing and devote their efforts to eliminating flaws in Microsoft's Windows 2000 and Windows XP operating systems.

Before beginning this task, Windows programmers and managers were sent to a day of training that focused on eliminating common coding mistakes and improving the program-design process to reduce errors. Every block of code was assigned an "owner" who was responsible for reviewing the design and the code. It took ten weeks to complete the review. Some bugs were identified and removed from source code; other bugs were scheduled for removal in subsequent releases. The Windows division manager committed that in the future, no software would ship containing serious, top-priority bugs.

Microsoft's internal quality-improvement program is called the Trustworthy Computing Security initiative. It was activated primarily by customer criticism about the lack of security in commercial software. Many complain that it is becoming too difficult for end users to keep current with all the software patches; they want to see software designed with security.

Although it is the world's biggest software manufacturer, Microsoft is known for rushing feature-packed software to market—and promising bug fixes in new releases. Microsoft has a lot of work to do to transform its deeply ingrained, highly bureaucratic culture to place a greater emphasis on software quality. Industry experts doubt that a day of training and a few weeks of bug fixing will have a lasting effect. They want to see Microsoft develop a plan to make systemic changes involving other key products, including Office, SQL Server, and Exchange.

Obviously, Microsoft is not the only software manufacturer guilty of turning out software that contains flaws. The National Infrastructure Protection Center's 2001 summary of software vulnerabilities is 70 pages long, representing dozens of well-known computer companies.

Sources: adapted from Byron Acohido, "Microsoft Does Security Sweep; Workers Drop Everything to Check Windows for Holes," *USA Today*, February 11, 2002 accessed at www.usatoday.com; Rick Perera, "Update: Gates Wants Security Top Priority at Microsoft," *Computerworld*, January 17, 2002, accessed at www.computerworld.com; Carol Silva, "Q&A Part II: Microsoft's Brian Valentine On Security," *Computerworld*, May 7, 2002, accessed at www.computerworld.com.

As you read this chapter, consider the following questions:

1. Why is there a need for quality software in business systems, industrial process control systems, and consumer products?

2. What are the ethical issues that software manufacturers face in making trade-offs between project schedules, project costs, and software quality?

3. What are the four most common types of software product liability claims and what actions must plaintiffs and defendants take to be successful?

4. What are the essential components of a software development methodology and what benefits can be derived from its use?

5. How can the Capability Maturity Model improve an organization's software development process?

6. What is a "safety-critical" system and what specific actions must be taken during the development of such a system?

STRATEGIES TO ENGINEER QUALITY SOFTWARE

High-quality software systems are easy to learn and easy to use. They perform the functions needed to meet their users' needs in a quick and efficient manner. They operate in a safe and dependable way and have a high degree of availability so that unexpected system downtime is kept to a minimum. Such software has long been required to support air traffic control systems, nuclear power plant operations, automobile safety systems, health care systems, military and defense systems, and space exploration technology. Now that the use of computers and software has become an integral part of many of our lives, more and more users are demanding high quality in the software they use every day. They can not afford to use software that results in crashed systems, lost work, lower productivity, and security holes through which intruders can spread viruses, steal data, or shut down Web sites. Software manufacturers are struggling with economic, ethical, and organizational issues associated with improving the quality of their software. This chapter will cover many of these issues.

A **software defect** is any error that, if not removed, would cause a software system to fail to meet the needs of its users. The impact of these defects can range from trivial (the computerized sensor in an ice cube maker on a refrigerator fails to recognize the ice cube tray is full and continues to make ice) to the tragic (the computerized control system for an automobile's antilock brakes malfunctions and sends the car into an uncontrollable spin). The defect may be very subtle and undetectable (a tax preparation software package miscalculates the amount of tax owed by a few dollars) or glaringly obvious (a payroll program generates checks with no deductions for social security or other taxes).

Software quality is the degree to which the attributes of a software product enable it to meet the needs of its users. **Quality management** addresses how to define, measure, and refine the quality of the information systems development process and the products (such as statements of requirements, flowcharts, and user documentation) developed during the various stages of the process. These products are known as **deliverables**. The objective of quality management is to help systems developers deliver high-quality systems—systems that meet the needs of their users. Unfortunately, it is the very rare exception that the first release of any software fully meets its users' expectations. Indeed, it is the rule, rather than the exception, that a software product does not work as well as its users would like until it has been used, some part of it has been found lacking, and the software has been corrected or upgraded.

A primary cause for poor software quality is that developers do not know how to or do not take the time to design quality into software from the very start. To do so, software developers must define and follow a set of rigorous, engineering-like principles and be committed to learn from past mistakes. In addition, they must take into account the environment within which their systems will operate and design systems that are as immune to human error as possible.

Even if software is well-designed, programmers make mistakes in turning design specifications into lines of code. It is estimated that an experienced programmer unknowingly injects about one defect into every ten lines of code. The programmers aren't incompetent or lazy—they're just human. All humans make mistakes, but in software, these mistakes result in defects.[2] The Microsoft Windows operating system is composed of hundreds of millions of lines of code bundled with products made by other Microsoft divisions. Even if 99.9 percent of those defects were identified and fixed before the product was released to the public, there would still be about one bug per 10,000 lines of code. Thus, it is highly likely that there are thousands of bugs in this software used daily by workers worldwide.

A third factor causing poor-quality software is that software manufacturers are under extreme pressure to reduce the time-to-market of their products. They are driven by the need to beat the competition in delivering new functionality to end users. They are also driven by the need to begin generating revenue to recover the cost of development of the product and to show a profit for shareholders. The resources and time budgeted to ensure the quality of the product are often cut under the intense pressure to get the new product shipped. When forced to make a choice between adding more end-user features or doing more software testing, most software development managers decide in favor of adding more features. After all, they reason, the defects can always be patched in the next release and that will give the customers an automatic incentive to want to upgrade. The additional features in this release will make it more useful and therefore easier to sell to customers. Although customers (who are stakeholders key to the success of the software manufacturer) may benefit from the new features added to a product, they also bear the burden of errors that aren't caught or fixed during testing and finalization of the software. Thus, many customers challenge whether the decision to "cut quality" in favor of feature enhancement is ethical.

As a result of the lack of quality in software, many organizations avoid buying the first release of a major software product or prohibit its use in mission-critical systems. Their rationale is that the first release will have many defects that will cause user problems. Because of the many defects in the first two popular Microsoft operating systems for end users (DOS and Windows) and their tendency to crash unexpectedly, many believe that Microsoft did not have a truly reliable operating system until its third major variation—Windows NT. Even software products that have become reliable over a long period of use will often falter unexpectedly (and at the worst possible time) when the operating conditions change. For instance, the software in the Cincinnati Bell phone switch had been thoroughly tested and had operated successfully for months after it was deployed in 1985. However, on a Sunday in October that same year, when the time changed from Daylight Savings to Standard Time, the switch failed because it was overwhelmed by the number of calls to the "time" phone number by people wanting to set their clocks. The volume of simultaneous calls to the same number was a change in operating conditions that no one had anticipated and for which no one had prepared.

The Importance of Software Quality

Most people think of business information systems when they first think about software. **Business information systems** are composed of a set of interrelated components including hardware, software, databases, networks, people, and procedures that collect data, process it, and disseminate the output. One common form of business system is the transaction-processing system used to capture and record business transactions. An example is a manufacturing organization's order-processing system that captures order information, processes it to update inventory and accounts receivable, and ensures that the order is filled and shipped on time to the customer. Another example is a bank's electronic-funds transfer system that is used to move funds from one bank to another. Yet a third is an airline's online reservation system used to reserve seats for passengers. The accurate, thorough, and timely processing of business transactions is a key requirement for such systems. The impact of a software defect in a transaction processing system can be devastating and result in upset customers, the eventual loss of customers, and a reduction in revenue. How many times would bank customers tolerate their funds being transferred to the wrong account before they would cease to do business with that bank?

Another type of business system is the **decision support system (DSS)**, which is used to improve decision-making effectiveness. The use of a DSS to develop accurate forecasts of customer demand, to recommend stocks and bonds for an investment portfolio, and to schedule shift workers to minimize cost while meeting customer service goals are examples of the use of such a system. If an incorrect business decision is made as a result of a software defect in the DSS, the impact can be just as devastating to an organization or its customers as an error in a transaction processing system.

Software is also used to control many industrial processes in an effort to reduce costs, eliminate human error, improve quality, and shorten the time it takes to make products. For example, steel manufacturers use process-control software to capture data from sensors about the equipment that rolls steel into bars and about the furnace used to heat the steel before it's rolled. Without process-control computers, workers could react to defects only after the fact and would have to guess at what adjustments should be made to the process to correct them. Process-control computers enable the process to be controlled and monitored for variations from operating standards to eliminate product defects *before* they can be made. Any defect in process-control software can lead to a decrease in product quality and an increase in waste and costs. It is even possible that a software defect can lead to unsafe operating conditions for employees.

Software is also used to control the operation of many industrial and consumer products—automobiles, medical diagnostic and treatment equipment, televisions, radios, stereos, refrigerators, stoves, washers, dryers, toasters, and irons. A software defect can result in relatively minor consequences, such as clothes not getting dry enough. A software defect can also result in much more serious consequences, such as a patient getting too strong an exposure to x-rays.

As a result of increasing utilization of computers and software in business, many companies are now in the software business whether they like it or not. The quality of software, its usability, and its timely development are critical to just about everything businesses do. The speed with which an organization develops software (whether it does it in-house or contracts it out) can put it ahead of or behind its competitors. Software problems may have been frustrating in the past, but mismanaged software can now be fatal to a business, causing it to miss product delivery dates, increase product development costs, and deliver products that customers perceive as having low quality.[3]

Business executives must struggle with ethical questions involving how much effort and money should be invested to assure high-quality software. A manager taking a short-term, profit-oriented view of the situation may feel that each additional day spent on quality assurance delays the release and sales revenues associated with a new product and that each additional dollar spent on quality assurance subtracts from the product's profit. A different manager may feel that it would be unethical to not fix all known problems before putting a product on the market and asking people to spend money on it.

Another key question for executives is whether their products could potentially cause damage and what their exposure would be if they did. Fortunately, software defects are rarely lethal and the number of personal injuries related to software failures is currently quite low. However, the use of software does introduce a new dimension to product liability that concerns many executives. Read the Legal Overview special interest box to learn more about software product liability.

LEGAL OVERVIEW

SOFTWARE PRODUCT LIABILITY

Software product litigation is certainly not new. One lawsuit from the early 1990s involved a financial institution that became insolvent because defects in a purchased software application caused errors in its integrated general ledger system, and in customers' passbooks and loan statements. These errors resulted in dissatisfied depositors withdrawing more than $5 million.[4] In a 1992 case involving an automobile manufacturer, a truck stalled because of a defect in the software that controlled the fuel injector of the truck. In the ensuing accident, a young child was killed. A state supreme court justified an award of $7.5 million in punitive damages against the manufacturer.[5]

The liability of manufacturers, sellers, lessors, and others for injuries caused by defective products is commonly referred to as **product liability**. There is no federal product liability law; product liability law is found mainly in common law (made by state judges) and in Article 2 of the Uniform Commercial Code that deals with the sale of goods.

If a software product defect causes injury to purchasers, lessees, or users of the product, the injured parties may be able to sue as a result. The injuries could include the loss of revenue or an increase in expenses due to a business disruption caused by the failure of the software to operate properly. Injuries could also include human injury or loss of life. Software product liability claims are frequently based on strict liability, negligence, breach of warranty, or misrepresentation.

Strict liability means that the defendant is held responsible for injuring another person, regardless of negligence or intent. The plaintiff must prove only

(continued)

that the software product is defective or unreasonably dangerous and that the defect caused the injury. There is no requirement to prove that the manufacturer was careless or negligent. Nor is there is a requirement to prove who caused the product to become defective. Indeed, all parties in the chain of distribution—the manufacturer, the subcontractors, and the distributors—are strictly liable for the injuries caused by the product and may be sued.

Defendants against a strict liability action may use several legal defenses, including the doctrine of supervening event, the government contractor defense, and an expired statute of limitations. Under the doctrine of supervening event, the original seller is not liable if the software is materially altered after it leaves the seller's possession it is the alteration that causes the injury. To establish the government contractor defense, a government software contractor must prove that 1) the precise specifications for the software were provided by the government, 2) the software conformed to these specifications, and 3) the contractor warned the government of any known defects in the software. Lastly, there are also statutes of limitations for claims of liability, so an injured party must file suit within a certain time period after the injury occurs.

When sued for **negligence**, a software supplier is not being held responsible for every product defect that causes customer or third-party loss. Instead, responsibility is limited to just those harmful defects that could have been detected and corrected through "reasonable" software development practices. Even where a contract may be written expressly to protect against supplier negligence, courts may disregard such terms in the contract as being unreasonable. Negligence is an area of great risk for software manufacturers or organizations with software-intensive products.

The defendant in a negligence case may either answer the charge with a legal justification for the alleged misconduct or demonstrate that the plaintiffs' own actions contributed at least in part to their injuries (**contributory negligence**). If proved, the defense of contributory negligence can reduce or totally eliminate the amount of damages the plaintiffs will receive. For example, if an individual uses a pair of pruning shears to trim his fingernails and ends up cutting off the tip of his finger, the defendant in this case could claim contributory negligence.

A **warranty** assures buyers or lessees that a product meets certain standards of quality. A warranty of quality may be either expressly stated or implied by law. Express warranties can be oral, written, or inferred from the seller's conduct. For example, sales contracts contain an implied warranty of merchantability. This implied warranty requires that the following standards be met: 1) the goods must be fit for the ordinary purpose for which they are used; 2) the goods must be adequately contained, packaged, and labeled; 3) the goods must be of an even kind, quality, and quantity within each unit; 4) the goods must conform to any promise or affirmation of fact made on the container or label; 5) the quality of the goods must pass without objection in the trade; and 6) the goods must meet a fair average or middle range of quality.[6] If the product fails to meet its warranty, the buyer or lessee can sue the seller or lessor for **breach of warranty**. Of course, most dissatisfied customers will first seek a replacement, a substitute product, or a refund before filing a lawsuit.

Software suppliers frequently write warranties in such a way as to limit their liability in the event of nonperformance. Although the software is warranted to run on a given machine configuration, no assurance is given as to what that software will do. Even if the contract specifically excludes the commitment of merchantability and fitness for a specific use, the court may find such a warranty disclaimer clause to be unreasonable and refuse to enforce the clause or refuse to enforce the entire contract. In determining whether warranty disclaimers are unreasonable, the court attempts to evaluate if the contract was made between two "equals" or between an expert and a novice. The relative education, experience, and bargaining power of the parties and whether the sales contract was offered on a take-it-or-leave-it basis are considered in making this determination.

The plaintiff must have a valid contract that the supplier did not fulfill in order to win a breach of warranty claim. Because the software supplier writes the warranty, this can be extremely difficult to prove. For example, in 1993, M.A. Mortenson Company had a new version of bid-preparation software installed for use by its estimators. During the course of preparing one new bid, the software allegedly malfunctioned several times, each time displaying the same cryptic error message. Nevertheless, the estimator submitted the resulting bid and Mortenson won the contract. Afterwards, Mortenson discovered that the bid was $1.95 million lower than intended and filed a breach-of-warranty suit against Timberline Software, makers of the bid software. Timberline acknowledged the existence of the bug, but the courts all ruled in favor of Timberline, ruling that the license agreement that came with the software explicitly barred recovery of the losses claimed by Mortenson.[7] Even if breach of warranty can be proven, the damages are generally limited to the amount of money paid for the product.

Intentional misrepresentation occurs when a seller or lessor either misrepresents the quality of a product or conceals a defect in it. A buyer or lessee who is injured because of intentional misrepresentation (for example, a cleaning product is labeled with "can be safely used in confined areas" and the user passes out from the fumes) can sue the seller for intentional misrepresentation or fraud. Advertising, salespersons' comments, invoices, and shipping labels are all forms of representation. Most software manufacturers use limited warranties and disclaimers to avoid any claim of misrepresentation.

Software Development Process

Developing information system software is not a simple process. It requires completing many activities that are complex themselves, with many dependencies among the various activities. There are system analysts, programmers, architects, database specialists, project managers, documentation specialists, trainers, and testers all involved in large projects. Each of these individuals has a role to play, and specific responsibilities and tasks. In addition, each will have decisions to make that can affect the quality of the software and the ability of users to use it effectively.

Many organizations for which the development of high-quality software is important have adopted a standard, proven work process (or **software development methodology**) that enables systems analysts, programmers, project managers, and others to make controlled and orderly progress in developing high-quality software. It defines the activities in the software development process and the individual and group responsibilities for accomplishing these activities. A methodology also recommends specific techniques for accomplishing the various activities (such as the use of a flowchart to document the logic of a computer program). A methodology also offers guidelines for managing the quality of the products produced during the various stages of the development life cycle. If an organization has developed such a methodology, it will be applied to any software development process that the company undertakes.

Like most things in life, it is far safer and cheaper to avoid software problems than to attempt to fix the damages after the fact. Studies at IBM, TRW, and GTE have shown that the cost to identify and remove a defect in the early stage of software development can be as much as 100 times less than removing a defect in an operating piece of software that has been distributed to hundreds or thousands of customers.[8] This increase in cost is due to two reasons. First, if a defect is uncovered in a later stage of the software development process, some rework of the deliverables produced in preceding stages must be done. Second, the later the error is detected, the greater the number of people affected by the error. In the extreme, consider the cost to communicate the details of a defect, distribute and apply software fixes, and possibly retrain end users for a software product that has been sold to hundreds or thousands of customers. Thus, most software development organizations try to identify and remove errors early in the software development process as a cost-saving measure and as the most efficient way to improve software quality.

Products containing inherent defects that cause harm to a user of the product, or to someone to whom the product was loaned or given, are the subjects of product liability suits. The use of an effective software development methodology can protect software manufacturers from legal liability for defective software in two ways. First, use of an effective software development methodology reduces the number of software errors that might occur. Second, if an organization follows a widely accepted software development methodology, it will be more difficult to prove negligence on the part of that company. However, even a *successful* defense against a product liability case can cost hundreds of thousands of dollars in legal fees. Thus, failure to practice reasonable and consistent software development can be quite serious in terms of liability exposure.

Software quality assurance refers to those methods within the software development methodology that are used to guarantee that the software being developed will operate reliably. Ideally, these methods are applied throughout the development life-cycle of the software product. At each stage in the software development process, appropriate quality assurance methods should be applied. However, some software manufacturing organizations that haven't established a formal, standard approach to quality assurance consider testing to be their only quality assurance method. Instead of checking for errors throughout the software development process, they rely primarily on testing just before the product ships to ensure some degree of quality.

There are different types of tests used in the software development process. We will discuss each in turn.

Dynamic

Software is developed in units called subroutines or programs. These units, in turn, are combined to form large systems. When a programmer completes a unit of software, one quality assurance measure is to test the code by actually entering test data and comparing the actual results to the expected results. This is called **dynamic testing**. There are two forms of dynamic testing: one based upon white-box testing and another based upon black-box testing:

- **Black-box testing** involves viewing the software unit as a device that has expected input and output behaviors but whose internal workings are unknown (a black box). If the unit demonstrates the expected behaviors for all the input data in the test suite, it then passes the test. Black-box testing takes place without the tester having any knowledge of the structure or nature of the actual code. For this reason, it is often done by someone other than the person who wrote the code.
- **White-box testing** treats the software unit as a device that has expected input and output behaviors but whose internal workings, unlike the unit in black-box testing, are known. White-box testing involves testing all possible logic paths through the software unit and is done with thorough knowledge of the logic of the software unit. The test data must be carefully constructed to cause each program statement to execute at least once. For example, if you wrote a program to calculate an employee's gross pay, you would develop test data to test cases in which the employee worked less than 40 hours and other cases in which he or she worked more than 40 hours (to check out the calculation of overtime pay).

Static Testing

In another form of testing, called **static testing**, special programs, called static analyzers, are run against the new code. Rather than reviewing input and output, the static analyzer is software that looks for suspicious patterns in programs that might indicate a defect.

Integration Testing

After successful unit testing, the software units are combined into an integrated subsystem that undergoes rigorous **integration testing**. This testing ensures that all the linkages among the various subsystems work successfully.

System Testing

After successful integration testing, the various subsystems are combined and **system testing** is conducted to test the entire system as a complete entity.

User Acceptance Testing

User acceptance testing is an independent test performed by trained end users to ensure that the system operates as expected from their viewpoints.

Capability Maturity Model® for Software

The Software Engineering Institute at Carnegie Mellon Institute developed a **Capability Maturity Model® for Software** (SW-CMM) in 1986. The SW-CMM is a globally used method to improve the capability and maturity of a software development organization. Its use can improve an organization's ability to predict and control quality, schedule, cost, cycle time, and productivity when acquiring, building, or enhancing software systems. It also helps software engineers to analyze, predict, and control selected properties of software systems. Identification of an organization's current maturity level enables it to identify specific actions to improve its future performance. Use of the model also enables the organization to track, evaluate, and demonstrate its progress as it improves over the years.

The SW-CMM defines five levels of software development process maturity (see Table 7.1) and identifies the issues most critical to software quality and process improvement. As the maturity level increases, the ability of the organization to deliver quality software on time and on budget is improved. In the awarding of software contracts, particularly with the government, organizations bidding on the contract may be required to have adopted the SW-CMM and be performing at a certain level.

After an organization decides to adopt the SW-CMM, it must conduct an assessment of its current software development practices (often using outside resources to ensure objectivity) and determine where they fit in the capability model. The assessment will identify specific opportunities for improvement and the action plans needed to upgrade the software development process. Over the course of a few years, the organization and software engineers will slowly raise their performances to the next level by identifying and completing actions required to improve. For example, Tanning Technology is a global systems integrator that took one year to move from level 2 to level 3.[9] SignalTree Solutions, another global IT outsourcing provider, is one of a few organizations to achieve level 5.[10] Table 7-1 shows the breakdown of the 1152 organizations whose maturity levels were assessed between 1997 and December 2001.[11]

Table 7-1 Software capability maturity levels

Maturity Level	Description	Percentage of Companies Assessed to be at This Level as of December 2001
1—Performed	The software development process is not standard and frequently fails to meet quality, cost, and schedule objectives.	25%
2—Managed	The software development process is planned, performed, monitored, and controlled; it frequently achieves the desired cost, schedule, and quality objectives.	40%
3—Defined	The software development process is managed and any deviations (beyond those allowed by customization guidelines) are documented, justified, reviewed, and approved.	24%
4—Quantitatively Managed	The software development process is controlled using statistical and other quantitative techniques throughout the duration of the project.	6%
5—Optimizing	The software development process is continuously improved through both incremental and innovative improvements.	7%

Source: Carnegie Mellon Software Engineering Institute, "CMMI[SM] for Systems Engineering/Software Engineering, Version 1.02," November 2000, pp. 26–27, accessed at www.sei.cmu.edu.

KEY ISSUES

Although defects in any system can cause serious problems, there are certain systems in which the consequences of software defects can be deadly. Here the stakes involved when creating quality software have been raised to the highest possible level. The ethical decisions involving a tradeoff—if one might even be considered—between quality and any other factor such as cost, ease of use, or time to market require extremely serious examination. The next sections will discuss safety-critical systems and the special precautions one must take in developing them.

Development of Safety-Critical Systems

A **safety-critical system** is one whose failure may cause injury or death to human beings. There are many safety-critical systems whose safe operation relies on the flawless performance of software—automobile anti-lock breaking systems, nuclear power plant reactor control systems, airplane navigation and control systems, roller coaster braking systems, elevator control systems, and numerous medical devices, to name just a few. The process of building software for such systems requires the use of highly trained professionals, formal and rigorous methods, and state-of-the-art tools. Failure to take strong measures to identify and remove software errors from safety-critical systems "is at best unprofessional and at worst can lead to disastrous consequences."[12] However, even with all these precautions, the software associated with safety-critical systems is still vulnerable to errors that can lead to injuries or the loss of human life.

In June 1994, a Chinook helicopter took off from Northern Ireland with 25 British Intelligence officials traveling to a security conference in Inverness. Just 18 minutes into its flight, the helicopter crashed on the peninsula of Kintyre in Argyll, Scotland, with no survivors. A handwritten memo by a senior Ministry of Defense procurement officer, dated January 11, 1995, revealed there were problems with the Chinook, "many of which were traced eventually back to software design and systems integration problems which were experienced from February–July 1994." In particular, the engine management software, which controlled the acceleration and deceleration of the engines, was suspect.[13]

One of the most widely cited software-related accidents in safety-critical systems involved a computerized radiation therapy machine called the Therac-25. This medical linear accelerator was designed to deliver either protons or electrons at various energy levels to create high-energy beams that would destroy tumors with minimal impact on the surrounding healthy tissue. Between June 1985 and January 1987, six known accidents involving massive overdoses were caused by software errors in the Therac-25, causing serious injuries and, in some cases, death.[14] The Therac bug was subtle. If a fast-typing operator mistakenly selected X-ray mode, and then used a particular editing key to change to electron mode, the display would appear to show a proper setting, when in reality the Therac had been configured to focus electrons at full power to a tiny spot on the body.[15]

The Therac-25 case illustrates that accidents seldom have a single root cause and that if only the symptoms of a problem are fixed, future accidents may still occur. With the benefit of 20-20 hindsight, there were a number of poor decisions made in regards to the development and use of the Therac-25 machine. The vendor decided to eliminate the hardware backup safety mechanisms present in earlier versions of the machine. The vendor elected to reuse parts of the Therac-20 software in the similar, but different Therac-25 machine. The software was poorly documented, and incomplete system testing was performed. The vendor failed to take a lead role in investigating the initial accident, insisting that the cause must have been human error. Lastly, hospitals did not push hard enough for answers and continued to use the Therac-25 machine even when they knew there had been accidents.

In the development of safety-critical systems, a key assumption must be that safety will *not* automatically result from following your organization's standard software development methodology. Safety-critical software must go through a much more rigorous and time-consuming development process than other kinds of software. All tasks, including requirements definition, systems analysis, design, coding, fault analysis, testing, implementation, and change control, require additional steps, more thorough documentation, and more checking and rechecking. As a result, safety-critical software takes much longer to complete and is much more expensive.

The key to getting this additional work done is the appointment of a project safety engineer who has explicit responsibility for the safety aspects of the system. The safety engineer uses a hazard logging and monitoring system to track hazards from project start to finish. The hazard log is used at each stage of the software development process to assess how that development stage has taken any hazards into account. Safety reviews are held throughout the development process. A robust configuration management system is used to track all safety-related documentation and to keep it consistent with the associated technical documentation. Informal documentation and verification is not acceptable for safety-critical systems development. Formal documentation and verification with sign-off signatures are required.

These additional activities and reviews mean that safety-critical software takes much longer to complete and is much more expensive. As a result, software developers are drawn into tough ethical dilemmas. For example, the use of hardware mechanisms to back up or verify critical software functions can help ensure safe operation and makes the software less critical. However, the use of such hardware may make the final product more expensive to manufacture or harder for the user to operate and thus make it less attractive than a competitor's product. These issues must be weighed carefully in order for the company to produce the safest product possible that will also be appealing to customers. Another key issue is deciding when enough quality assurance and testing activities have been performed. How much is enough when you are building a product that can cause loss of human life? At some point, software developers must sign off on the fact that sufficient quality assurance activities have been completed; determining what activities are sufficient demands careful decision-making.

In designing, building, and operating a safety-critical system, a great deal of effort must go into considering what can go wrong, how likely it is to happen, what are the consequences if it happens, and how can the risks be averted or mitigated. One approach to answer these questions is through a formal risk analysis. **Risk** is the probability of an undesirable event occurring times the magnitude of the consequence of the event happening (damage to property, loss of money, injury to people, loss of life, and so on). For example, if an undesirable event has a 1 percent probability of occurring and its consequences are quantified as costing $1,000,000, then the risk can be determined to be .01 × $1,000,000, or $10,000. This risk would be considered greater than an event with a 10 percent probability of occurring and a cost of $100 (.10 × $100 = $10). Risk analyses are key for safety-critical systems, but are useful for all kinds of software development as well.

Redundancy is the provision of multiple interchangeable components to perform a single function in order to cope with failures and errors. A simple redundant system would be an automobile with a spare tire or a parachute with a backup chute attached. A more complex redundant system used in IT would be the redundant array of independent disks (RAID), commonly used in high-volume data storage for file servers. RAID systems use many small-capacity disk drives to store large

175

amounts of data and to provide increased reliability and redundancy. Should one of the drives fail, it can be removed and a new one inserted in its place. Data on the failed disk is rebuilt automatically without the server ever having to be shut down.

As with other systems, improper planning for redundant systems can lead to unexpected problems. In March 2000, a subcontractor laying new communications lines in Eagan, Minnesota, accidentally bored through clusters of existing cables, cutting both the primary and backup lines that link Northwest Airline's Minneapolis-St. Paul hub to the rest of the nation, leaving thousands of airline passengers stranded nationwide. Airline officials were quite disturbed to learn that the primary and back-up communications lines were essentially next to one another. In April 2001, a power failure in northern Virginia caused AOL's primary power supply to fail. At the same time and for other reasons, one of AOL's backup power supplies failed, briefly denying AOL's 28 million users access to its Instant Messenger service.

After all risks pertinent to a system are determined, it is necessary to decide what level of risk is acceptable. This is an extremely difficult and controversial decision because it involves forming personal judgments about the value of human life, assessing potential liability in case of an accident, evaluating the surrounding natural environment, and estimating the costs and benefits of the system. System modifications must be made if the level of risk inherent in the design is judged to be too great. The modifications can include adding redundant components or using safety shut down systems, containment vessels, protective walls, or escape systems. Another approach is to mitigate the consequences of failure by devising emergency procedures and evacuation plans. In all cases, they must decide that if human life is at stake, how safe is safe enough?

Manufacturers of safety-critical systems must sometimes make a decision about whether to recall a product when data indicates there may be a problem. For example, automobile manufacturers have been known to weigh the cost of potential future lawsuits versus the cost of an automobile recall in deciding whether to issue a recall. On the other hand, drivers and passengers in those automobiles (and in many cases, the courts) did not find this to be an ethical approach to decision-making. Manufacturers of medical equipment and airplanes have had to make similar decisions. Such a decision can be complicated when the data cannot pinpoint the cause of the problem. For example, in 2000, there was great controversy over the use of Firestone tires on Ford Explorers; the incidence of tires blowing out and Explorers rolling over had caused multiple injuries and fatalities. However, it was difficult to determine if the cause of the tipping was poor automobile design, faulty tires, or incorrectly inflated tires. Consumers' confidence in both products was shaken.

Reliability is the probability of a component or system performing its mission over a certain length of time. If a component has a reliability of 99.9 percent, it has one chance in one thousand of failing over the life of its mission. Although this may seem very low, remember that most systems are made up of many components. In such a case, the overall reliability of the complete system is based upon statistical probabilities. For example, imagine that you are building a complex system made up of several components, each with 99.9 percent reliability. As you add more components to the system, the system becomes more complex and the chance of a failure increases. With a system of seven components and no redundancy built in, you have a 93.8 percent probability of operating successfully with no component malfunctions over the life of the mission. If you build a system using ten components with 99.9 percent reliability, the probability of no failure falls to less than 60 percent! Thus, building redundancy into systems that are both complex (made up of many components) and safety-critical is imperative.

One of the most important and difficult areas of safety-critical system design is the human-system interface. Human behavior is not nearly as predictable as the reliability of the hardware and software components of a complex system. The system designer must consider what human operators might do to cause the system to operate less safely or effectively. The challenge is to design a system so that it will not only operate as it should, but also leave the operator little room for erroneous judgment. For instance, it is important for a self-medicating pain relief system to allow a patient to press a button to receive more pain reliever, but it is also important for the machine to regulate itself so that an overdose is not given. Additional risk can be introduced if a designer does not anticipate the information an operator needs and how that operator will react under the daily pressures of actual operation, especially when in a crisis situation. Some people keep their wits about them and perform admirably in an emergency, but others may panic and only make a bad situation worse.

Poor human-system interface design can greatly increase risk, sometimes with tragic consequences. For example, in July 1988, the guided missile cruiser *U.S.S. Vincennes* mistook an Iranian Air commercial flight with 290 people on board for an enemy F-14 and shot down the airliner over international waters in the Persian Gulf. Some investigators blame the tragedy on the confusing human-system interface of the $500 million Aegis radar and weapons control system. The Aegis radar on the *U.S.S. Vincennes* locked onto an Airbus 300, but it was misidentified as a much smaller F-14 jet fighter by its human operators. The Aegis operators also misinterpreted the system signals and thought that the target was descending even though the airbus was, in fact, climbing. A third human error was made in determining the target altitude—it was wrong by 4,000 feet. As a result of this combination of human errors, the Vincennes crew thought the ship was under attack and the Airbus was shot down.[16]

Quality Management Standards

The International Organization for Standardization (ISO), founded in 1947, is a worldwide federation of national standards bodies from some 100 countries. The ISO issued its 9000 series of business management standards in 1988. These standards require organizations to develop formal, quality management systems that focus on identifying and meeting the needs, wants, and expectations of their customers.

The **ISO 9000** standard serves many different industries and organizations as a guide to quality products, services, and management. There are approximately 350,000 ISO 9000-certified organizations in over 150 countries. Although companies can use the standard as a management guide for their own purposes in achieving effective control, the bottom line for many is having a certificate awarded by a qualified external agency to say that they have achieved ISO 9000 certification. Many companies and government agencies specify that a company must be ISO 9000 certified to win a contract.

In order to get this coveted certificate, it is necessary for the organization to submit to an examination by an external assessor. To be certified, a company must do three things:

1. Have written procedures for everything they do.
2. Follow those procedures.
3. Prove to an auditor that they have written procedures and that they follow those procedures. This can require observation of actual work practices and interviews with customers, suppliers, and employees.

The various ISO 9000 series of standards address the following activities:

ISO 9001	Design, development, production, installation, servicing
ISO 9002	Production, installation, servicing
ISO 9003	Inspection and testing
ISO 9000-3	The development, supply, and maintenance of software
ISO 9004	Quality management and quality systems elements

Failure Mode and Effects Analysis (FMEA) is an important technique used to develop any ISO 9000-compliant quality system. FMEA is used as a reliability evaluation technique to determine the effect of system and equipment failures. Failures are classified according to their impact on mission success, personnel safety, equipment safety, customer satisfaction, and customer safety. The goal of FMEA is to identify potential design and process failures early in the design process when they are relatively easy and inexpensive to correct. A failure mode describes the way in which a product or process could fail to perform its desired functions as described by the needs, wants, and expectations of the customer. An effect is an adverse consequence that the customer might experience. Unfortunately, most systems are so complex that there is seldom a one-to-one relationship between cause and effect. Instead, a single cause may have multiple effects and a combination of causes may lead to one effect or to multiple effects.

Another example of a quality management standard comes from the aviation industry. Over the past 25 years, the role of software in aircraft manufacturing has expanded from controlling simple measurement devices to being involved in almost all aircraft functions, including navigation, flight control, and cockpit management. Given the expanded use of software in the aviation industry, the industry and its regulatory community needed an effective means of evaluating safety-critical software. To meet this need, the Radio Technical Commission for Aeronautics (RTCA), a private, non-profit organization that develops recommendations for use by the Federal Aviation Administration (FAA) and the private sector, developed DO-178B/EUROCAE ED-128 as the evaluation standard for the international aviation community. Aviation software developers use the standard to achieve a high level of confidence in the safety-critical software.

Table 7-2 provides a useful checklist for an organization interested in upgrading the quality of the software that it produces. The preferred response to each question is "Yes."

Table 7-2   Manager's checklist for improving software quality

Questions	Yes	No
Has senior management made a commitment to quality software?	____	____
Have you used the Capability Maturity Model® for Software (SW-CMM) to evaluate your organization's software development process?	____	____
Have you adopted the use of a standard software development methodology?	____	____
Does the methodology place a heavy emphasis on quality management and address how to define, measure, and refine the quality of the software development process and the products developed during the various stages of the process?	____	____
Are software project managers and team members trained in the use of this methodology?	____	____
Are software project managers and team members held accountable for following this methodology?	____	____
Is there a strong effort made to identify and remove errors as early as possible in the software development process?	____	____
In the testing of software, is the use of both static and dynamic testing used?	____	____
Is white-box and black-box testing used?	____	____
Has an honest assessment been made to determine if the software being developed is safety-critical?	____	____
If the software is safety-critical, are additional tools and methods employed and do they include the following: project safety engineer, hazard logs, safety reviews, formal configuration management systems, rigorous documentation, risk analysis processes, and the FMEA technique?	____	____

Summary

1. Why is there a need for quality software in business systems, industrial process control systems, and consumer products?

High-quality software systems are needed because they are easy to learn and easy to use, perform the functions that meet users' needs in a quick and efficient manner, operate safely and dependably, and have a high degree of availability so that unexpected downtime is kept to a minimum.

Such high-quality software has long been required to support air traffic control systems, nuclear power plant operations, automobile safety systems, health care systems, military and defense systems, and space exploration technology. Now that the use of computers and software has become an integral part of many of our lives, more and more users are demanding high quality in their software. They cannot longer afford to use software that results in crashed systems, lost work, lower productivity, and security holes through which intruders can spread viruses, steal data, and shut down Web sites.

2. What are the ethical issues that software manufacturers face in making trade-offs between project schedule, project cost, and software quality?

Software manufacturers are under extreme pressure to reduce the time to market of their products. They are driven by the need to beat the competition in delivering new functionality to end users. They are also driven by the need to begin generating revenue to recover the cost of development of the product and to show a profit for shareholders. The resources and time budgeted to ensure the quality of the product are often cut under the intense pressure to get the new product shipped. When forced to make a choice between adding desirable end-user features or doing more software testing, most software development managers decide in favor of adding more features. After all, they reason, the defects can always be patched in the next release. This will give the customers an automatic incentive to upgrade. The additional features in this release will make it more useful and therefore easier to sell to customers. Many customers challenge whether the decision to "cut quality" is ethical.

3. What are the four most common types of software product liability claims and what actions must plaintiffs and defendants take to be successful?

Software product liability claims are frequently based on strict liability, negligence, breach of warranty, or misrepresentation. Strict liability means that the defendant is responsible for injuring another person regardless of negligence or intent. The plaintiff must prove only that the software product is defective or unreasonably dangerous and that the defect caused the injury. When sued for negligence, a software supplier is not being held responsible for every product defect that causes customer or third-party loss. Instead, responsibility is limited to just those harmful defects that could have been detected and corrected through "reasonable" software development practices. If the product fails to meet its warranty, the buyer or lessee can sue the seller or lessor for breach of warranty. The plaintiff must have a valid contract that the supplier did not fulfill in order to win a breach of warranty claim. Intentional misrepresentation occurs when a seller or lessor either misrepresents the quality of a product or conceals a defect in it. Most software manufacturers use limited warranties and disclaimers to avoid any claim of misrepresentation.

4. What are the essential components of a software development methodology and what benefits can be derived from its use?

A software development methodology defines the activities in the software development process and the individual and group responsibilities for accomplishing these activities; recommends specific techniques for accomplishing the activities; and offers guidelines for managing the quality of the products produced during the various stages of the software development life cycle.

The use of an effective software development methodology enables a software manufacturer to produce high-quality software, forecast project completion milestones, and reduce the overall cost to develop and support software. It also protects software manufacturers from legal liability for defective software in two ways: it reduces the number of software errors that could potentially cause damage and it makes it more difficult to prove negligence.

5. How can the Capability Maturity Model® for Software (SW-CMM) improve an organization's software development process?

The SW-CMM defines five levels of software development process maturity and identifies the issues most critical to software quality and process improvement. Its use can improve an organization's ability to predict and control quality, schedule, costs, cycle time, and productivity when acquiring, building, or enhancing software systems. It also helps software engineers to analyze, predict, and control selected properties of software systems.

6. What is a safety-critical system and what specific actions must be taken during the development of such a system?

A safety-critical system is one whose failure may cause injury or death to human beings. In the development of safety-critical systems, a key assumption is that safety will *not* automatically result from following an organization's standard software development methodology. Safety-critical software must go through a much more rigorous and time-consuming development and testing process than other kinds of software; the appointment of a project safety engineer and the use of a hazard log and risk analysis are common.

Review Questions

1. What is a software defect?
2. What is risk? What is reliability?
3. What must a defendant prove to win a strict liability claim? What legal defenses are commonly used by defendants?
4. What must a defendant prove to win a negligence claim? What legal defenses are commonly used by defendants?
5. What is breach of warranty? What must the plaintiff prove to win a breach of warranty claim? How can a software manufacturer defend against potential breach of warranty claims?
6. What is the difference between quality management and quality assurance?
7. What is a software development methodology? What is its purpose?
8. Why is it critical to identify and remove defects early in the software development process?
9. Identify the various types of testing used in software development.
10. What is a safety-critical system? What additional precautions must be taken during the development of such a system?
11. What is the purpose of the ISO 9000 quality standard? How is it similar to and different from the SW-CMM?
12. What is FMEA? What is its goal?

Discussion Questions

1. Discuss the relative importance of software quality for business information systems, industrial and consumer products, and industrial process control systems.
2. Explain why the cost to identify and remove a defect in the early stages of software development might be as much as 100 times less than removing a defect in a piece of software that has been distributed to hundreds of customers.
3. Why is it important for software manufacturers to follow a rigorous software development methodology?
4. Discuss the implications for a project team developing a piece of software that has been classified as safety-critical.
5. You have been asked to draft a boilerplate warranty for a software contractor that will absolutely protect the firm from being sued successfully for negligence or breach of contract. Is this possible? Why or why not?
6. Discuss why an organization may elect to use a separate, independent testing team rather than the group of people who originally developed the software to conduct quality tests.

What Would You Do?

1. Read the fictitious Killer Robot case found at the Web site for the Online Ethics Center for Engineering & Science at www.onlineethics.com (look under Computer Science and Internet cases). The case begins with the indictment of a programmer for manslaughter for writing faulty code that resulted in the death of a robot operator. Slowly, over the course of many articles, you are introduced to several factors within the corporation that contributed to the accident. Read this case and answer the following questions:

   a. Responsibility for an accident is rarely clearly defined and able to be traced to one or two individuals or causes. In this fictitious case, it is clear that a large number of people share responsibility for the accident. Identify all the people who you think were at least partially responsible for the death of Bart Mathews, and why you think so.

   b. Imagine that you are the leader of a task force assigned to correct the problems uncovered by this accident. Develop your "top-ten" list of actions that need to be taken to avoid future problems. What process would you use to identify the most critical actions?

   c. If you were in Ms. Yardley's position, what would you have done when Ray Johnson told you to fake the test results? How would you justify your decision?

2. Your manager insists that the chemical reactor shutdown control software to which you have been assigned is not safety-critical software. It is designed to sense temperatures and pressures within a giant, 50,000-gallon stainless steel vat and dump in chemical retardants that will slow down the chemical reaction if it gets out of control. In the worst possible scenario, failure to stop a runaway reaction would result in a large explosion that would send fragments of the stainless steel vat flying and spray caustic soda in all directions. He points out that the stainless steel vat is surrounded by two sets of protective concrete walls and that the reactor's human operators can intervene in case of a software failure. He feels that these measures ensure that the plant employees and the surrounding neighborhood would suffer no harmful effects if the shutdown software failed. Besides, he argues, the plant is already more than a year behind its scheduled start-up date. He cannot afford the additional time required

to develop the software if it is classified as safety-critical. How would you work with your manager and other appropriate resources to decide whether the software is safety-critical?

3. You are a senior software development consultant with a major consulting firm. It has been two years since you conducted the initial assessment of the ABCXYZ Corporation's software development process. Using the SW-CMM, you determined their level of maturity to be at level 1—Performed. The organization has spent a lot of time and effort following your recommended action steps to raise their level of process maturity to the next level, which is 2—Managed. They appointed a senior member of the IT organization to be a process management guru at a cost of $150,000 per year to lead them in their improvement effort. This individual adopted a standard software development methodology and forced all project managers to go through a one-week training course at a total cost of over $2 million.

Unfortunately, these efforts did not result in much real improvement in process maturity because senior management failed to hold project managers accountable for actually using the standard software development methodology on their projects. Too many project managers were able to convince senior management that the use of the new methodology was not really necessary on their project and would just slow things down. However, you are concerned that when senior management learns that no real progress has been made, they will refuse to accept partial blame for the failure and instead drop all attempts at further improvement. You want senior management to ensure that the new methodology is used on all projects—no more exceptions. What would you do?

4. You are the CEO for a small, struggling software firm that produces educational software for high-school students. Your latest software is designed to help students improve their SAT and ACT scores for getting into college. To prove the value of your software, a group of 50 students who had taken the ACT test were given the test a second time, after using your software for just two weeks. Unfortunately, there was no dramatic increase in their scores. A statistician hired to ensure objectivity in measuring the results has claimed the variation in test scores was statistically insignificant.

It will take at least six months for a small core group of educators and systems analysts to start over again from scratch and design a viable product. Then it could take another six months of programming and testing to complete the software. You could go ahead and release the current version of the product and then, when the new product is ready, announce it as a "new release." This would generate the cash flow necessary to keep your company afloat and save the jobs of 10 or more of your 15 employees.

Given this information about your company's product, what would you do?

Cases

1. Airbags—Safety Critical Systems in Automobiles

The National Highway Traffic Safety Administration (NHTSA) estimates that air bags have prevented nearly 3,500 deaths. However, most automobiles carry only one-size-fits-all air bags that are designed to protect an average-sized man. These air bags fire out of the dashboard at more than 140 miles per hour, unleashing too much energy for smaller people and children. As a result, over 100 children and small adults have been killed by airbags, and thousands have incurred nonfatal injuries.

The release of air bags is controlled by a highly sophisticated microprocessor. The decision to fire or not fire must be made within 20 milliseconds of impact or the bag won't be fully deployed before the occupant strikes the dash. During this instant of time, the control microprocessor of the air bag estimates the severity of the accident from accelerometer readings taken by an instrument mounted in front of the passenger's compartment. Then, based on its estimate of the severity of impact, the microprocessor decides whether to fire the air bag. Important parameters not considered in the microprocessor's determination process include the size of the passenger, whether the passenger is belted, and the location of the passenger within the automobile at the instant of impact.

The NHTSA has ordered vehicle makers to install smarter airbags in 25 percent of new passenger cars and light trucks by September 1, 2002, and in all new models by September 1, 2005. The NHTSA will allow the auto manufacturers to use different approaches rather than mandate the specific technologies to be used—as long as their vehicles pass a series of standard tests. The auto industry is working hard to meet the new standards and has considered a variety of approaches to doing so, including the following: 1) sensors that can feed the air bag controller details about vehicle occupants at the instant of impact, 2) airbags that are capable of inflating at several different rates and pressures, and 3) vehicles with the ability to prejudge the severity of a crash before it begins by using radar systems now being developed for collision avoidance.

Any changes in air bag technology must be coordinated with advanced seat belt features under development. Belt manufacturers have developed ways to draw belt webbing back into the reel as a crash begins so that the passenger is cinched more tightly against the seat. Engineers are also designing load limiters into seat belts to reduce belt tension so that it does not reach a level harmful to the occupants. Inflatable seat belts with shoulder straps that would puff up to hold and cushion occupants during a crash are also being considered.

Meanwhile, many industry advocates argue against making air bags more complex. They claim that the current air bag technology already has too many points of failure and is far from foolproof. To support this claim, they point out that millions of cars have been recalled due to problems with the airbags firing at random when no accident was taking place. They say that the best solution is education rather than additional technology and that people should be taught to use seat belts and to put children and small adults in the back seat rather than installing complex and costly equipment. In spite of such criticism, a number of new airbag-control ideas have already been implemented.

Siemens VDO Automotive, a supplier of electronics and electrical technologies to the automotive industry, has developed an integrated safety system incorporating a seat-occupancy detection system. This system includes a mat located under the cushion of the passenger seat measuring the passenger's weight and weight distribution and a three-dimensional camera that signals the passenger's position to the airbag control unit. These parameters are transmitted to the airbag controller where they are processed together with the results of the camera surveillance and the vehicle acceleration caused by the accident. Multistage, gas-generated inflators are used to vary the volume of the airbag upon deployment.

Another carmaker, Jaguar, developed an Adaptive Restraint Technology System to control passenger airbag deployment; the system came as standard in the Jaguar's 2001 XK Series sports cars. Four ultrasound sensors determine the position of the passenger's head and torso relative to the airbag deployment door. If the passenger is too close to the dash, the air bag is deactivated and a warning light comes on to indicate inactive status. After the passenger is sufficiently clear, the bag is made active and the light goes out. For the driver, a position sensor determines seat location relative to the steering wheel. On both driver and passenger seat belts, other sensors indicate to the system whether the seat belt is fastened. Within the vehicle structure, impact sensors on the front cross-member panel and on the sides of the car determine impact severity. A central processor monitors all sensor data. It then activates the seatbelt pre-tensioners and stages the air bags for full or partial inflation.

Questions:

1. Do you agree with the critics who argue against making airbags even more complex? Why or why not?

2. Imagine that you are the vice-president of manufacturing for an automobile manufacturer. What are the factors you would consider in reaching a decision on whether to put additional time and money into the improvement of airbags? At what point would you feel comfortable that your company is spending the right amount of money in protecting its customers?

3. Visit the NHTSA Web site and read about the auto industry's current progress toward meeting the new safety guidelines. Write a paragraph summarizing your findings.

Sources: adapted from Peter Weiss, "Curbing Air Bags' Dangerous Excesses," *Science News Online*, September 26, 1998; "Siemens Automatic 3-D Occupant Sensor Technology To See Production In 2002," Seimens News Release, March 2, 2001, accessed at http://siemens.de; "Siemens VDO Automotive Technology Offers New Dimensions In Occupant Sensing Optical Passenger Detection From Siemens VDO Automotive," Siemens News Release, September 12, 2001, accessed at http://siemens.de; John Lewis, Rick DeMeis, Darius Mehri, and Kevin Russelburg, "Hot New Auto Technologies for 2001," *Design News Online*, October 2, 2000, accessed at www.manufacturing.net; Jeff Green, "What's New on the Lot? A Safer Ride," *Business Week Online*, May 7, 2001.

2. Patriot Missile Failure

The Patriot is an Army surface-to-air missile system that defends against aircraft and cruise missiles, and more recently, against short-range ballistic missiles. Designed in the 1960s as an antiaircraft weapon, the Army enhanced the Patriot system in the 1980s to provide a limited-area defense against short-range ballistic missiles. This antimissile capability is incorporated into the Patriot PAC-2 version of the missile.

Following the Iraqi invasion of Kuwait in August 1990, the United States deployed the Patriot PAC-2 missile to Saudi Arabia during Operation Desert Shield. At the start of Desert Shield, there were only three Patriot PAC-2 missiles in the United States arsenal. PAC-2 production was accelerated so that by January 1991, 480 missiles were available. Patriot battalions were deployed to Saudi Arabia and then to Israel. These battalions were generally placed in permanent positions to defend key assets, military personnel, and citizens against Iraqi Scud missiles. Iraq launched 81 modified-Scud tactical ballistic missiles into Israel and Saudi Arabia during this conflict.

The Iraqis modified the Scud missile to increase its range and boost its speed by as much as 25 percent. This was done by reducing the weight of the warhead, enlarging the fuel tanks, and modifying its flight so that all of the fuel was burned during the early phase of flight (rather than continuously). As a result of these modifications, the Iraqi Scud missile was structurally unstable and often broke into pieces in the upper atmosphere. All this made it extremely difficult to intercept the warhead because its flight path was unpredictable and radar mistakenly could lock onto pieces of the fragmented missile rather than the warhead.

The Patriot missile is launched and guided to the target through three phases. First, the missile guidance system turns the Patriot launcher to face the incoming missile. Second, the Patriot's computer control system guides the missile toward the incoming Scud missile. Third, the Patriot missile's internal radar receiver guides it to intercept the incoming missile.

The Patriot's radar is constantly sweeping the skies for any airborne object with the flight characteristics of a Scud missile. The range gate is an electronic detection device within the radar system that uses the last observation of such an object to forecast an area in the air space where the radar system should next see the object, if it truly is a Scud. This forecast is a function of the object's observed velocity and the time of the last radar detection. If the range gate determines that the detected target is a Scud, and if the Scud is in the Patriot's firing range, the Patriot battery fires its missiles.

On February 11, 1991, the Patriot Project Office received data from Patriots deployed in Israel identifying a 20 percent shift in the Patriot system's radar range gate after the system had been running for eight consecutive hours. Such a shift is quite significant. It meant that the target was no longer in the center of the range gate and the probability of successfully tracking the target was greatly reduced. Further, it was known that the Patriot system could not track a Scud when there is a range gate shift of 50 percent or more.

In the Patriot radar system, time is measured by the system's internal clock in tenths of a second *from time of system startup*. The nature of the software problem was that the longer the system was running, the less accurate the elapsed time calculation. Consequently, after the Patriot computer control system has been running continuously for extended periods, the range gate makes an inaccurate estimate of the area in the air space where the radar system should next see the object. This in turn can cause the radar to lose the target and fool the system into thinking there is no incoming Scud.

Army officials assumed that other Patriot users were not running their systems for eight hours or more at a time and that the Israeli experience was an anomaly. However, the Patriot Project Office analyzed the Israeli data and confirmed some loss in targeting accuracy. As a result, they made a software change to compensate for the inaccurate time calculation. This change was included in a modified software version that was released on February 16, 1991.

The Patriot Project Office sent a message to Patriot users on February 21, 1991, informing them that very long run times could cause a shift in the range gate, resulting in difficulty tracking the target. The message also advised them that a software change was on the way that would improve the system's targeting. However, the message did not specify what constitutes "very long run times." Patriot Project Office officials presumed that the users would not continuously run the batteries for such extended periods of time that the Patriot would fail to track targets. Therefore, they did not think that more detailed guidance was required.

On February 25, 1991, a Patriot missile defense system operating at Dhahran, Saudi Arabia, failed to track and intercept an incoming Scud. The enemy missile subsequently hit an Army barracks and killed twenty-eight Americans. The ensuing investigation revealed that at the time of the incident, the Patriot battery had been operating continuously for over 100 hours. Because the system had been on so long, the resulting inaccuracy in the time calculation caused the range gate to shift so much that the system could not track and identify the incoming Scud. Consequently, the Patriot missile battery did not even engage the Scud missile.

By cruel fate, the very next day modified software that compensated for the inaccurate time calculation arrived in Dhahran. Army officials attributed the delay to the time it took to arrange air and ground transportation in a wartime environment.

The Army did not have the luxury of collecting definitive performance data during Operation Desert Storm. After all, they were operating in a war zone rather than on a test range. As a result, there is insufficient and conflicting data on the effectiveness of the Patriot missile. Lack of conclusive data did not deter the Army, officials, and the press from issuing their estimates, though, and the numbers vary widely. At one extreme is an early report that claimed the Patriot destroyed about 96 percent of the Scuds *engaged* in Saudi Arabia and Israel. (This presumably does not include those Scuds it failed to engage due to the software error.) At the other extreme, in only about 9 percent of the Patriot's Operation Desert Storm engagements did observers actually see a Scud destroyed or disabled after a Patriot detonated close to the Scud. Of course, some "kills" could have been effected out of their range of vision.

Questions:

1. With the benefit of 20-20 hindsight, what specific steps could have been taken during development of the Patriot software to avoid the operational problems that led to the loss of life? Do you think these steps would have improved the effectiveness of the Patriot system to the point that it would have been obvious to all that

it was a strong deterrent against the Scud missile? Why or why not?

2. What ethical decisions do you think United States military made in deciding to deploy the Patriot missile to Israel and Saudi Arabia and in the reporting of the effectiveness of this missile system?

3. What key lessons can be taken from this example of the development of safety-critical software and applied to the development of business information system software?

Sources: adapted from Ralph V. Carlone, "Patriot Missile Defense—Software Problem Led to Systems Failure at Dhahran, Saudi Arabia," GAO Report B-247094, February 24, 1992, accessed at www.fas.org; "Data Does Not Exist to Say Conclusively How Well Patriot Performed," GAO Report B-250335, September 22, 1992, accessed at www.fas.org; "Taiwan Interested in Latest Patriot Missiles," *Asian Political News*, November 30, 1998, accessed at www.findarticles.com; "U.S. To Sell 14 Upgraded Missile Systems to S. Korea," Asian Political News, November 15, 1999, accessed at www.findarticles.com; "Frontline: The Gulf War: Weapons: MIM -104 Patriot," January 20,1996, PBS Web site accessed at www.pbs.org.

Endnotes

1 Rick Perera, "Update: Gates Wants Security Top Priority at Microsoft," *Computerworld*, January 17, 2002, accessed at www.computerworld.com.

2 Watts S. Humphrey, "Why Quality Pays," *Computerworld*, May 20, 2002, accessed at www.computerworld.com.

3 Watts S. Humphrey, "Why Quality Pays," *Computerworld*, May 20, 2002, accessed at www.computerworld.com.

4 Bruce A. Bierhan, "Faulty Software or Puffery," *Computerworld*, March 28, 1994, accessed at www.computerworld.com.

5 Cem Kaner, "Bad Software—Who Is Liable?", Keynote Address: Quality Assurance Institute Regional Conference, Seattle, June 1998, accessed at www.badsoftware.com.

6 Henry R. Cheeseman, *Contemporary Business Law, 3rd Edition*, pg. 362, ©2000, Prentice-Hall, Inc., Upper Saddle River, NJ.

7 "Software Company Not Liable for $2 Million Bug In Bid," *Engineering Times*, July 2000, accessed at www.nspe.org.

8 Barry W. Boehm, *Software Engineering Economics*, Englewood Cliffs, NJ, Prentice-Hall, 1981.

9 Business Wire, "Tanning Technology India Awarded SEI SW-CMM Level 3 Certification," *Business Wire*, January 8, 2001, accessed at www.findarticles.com.

10 "SignalTree Solutions Furthers Position in Hospitality Sector with La Quinta Inns Contract Signing," *PRNewswire*, July 16, 2001, accessed at www.findarticles.com.

11 "Process Maturity Profile of the Software Community 2001 Year End Update," Carnegie Mellon Software Engineering Institute, March 2002, accessed at www.sei.cmu.edu.

12 Jonathan P. Bowen, "The Ethics of Safety-Critical Systems," accessed at www.cs.rdg.ac.uk on October 30, 2001.

13 Tony Collins, "Minister Denies Chinook Claim," *Computer Weekly*, July 1, 1999, accessed at www.computerweekly.com.

14 Nancy G. Leveson and Clark S. Turned, "An Investigation of the Therac-25 Accidents," *IEEE Computer*, Vol. 26, No. 7, pp. 18–41, July 1993.

15 "Getting Serious with Year 2000, Sometimes Bugs Can Be Deadly," *InfoWeek*, April 13, 1998, accessed at www.infoworld.com.

16 George J. Church, "High-Tech Horror," *Time*, July 18, 1988, accessed at www.time.com.

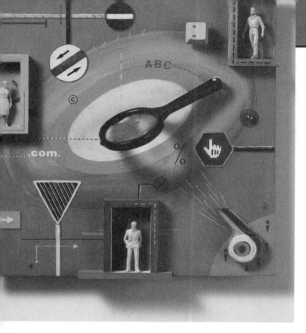

Chapter 8

Employer/Employee ISSUES

"The U.S. faces a worker gap and a skills gap—and both are right around the corner."
— James E. Oesterreicher, former chairman of J.C. Penney, who helped prepare the 2002 Aspen Institute study of the future labor market[1]

Global Crossing was at one time one of the largest owners of fiber optic cables, with a network linking more than 200 major cities in 27 countries. On January 28, 2002, it filed for Chapter 11 bankruptcy protection with approximately $12 billion in debt, and assets worth more than $22 billion. The next month, the firm announced more bad news—it would write off at least $8 billion of those assets. Holders of Global Crossing common shares of stock would likely lose their entire investments.

It is standard industry practice for telecommunications companies to fill in gaps in their fiber optic networks by leasing capacity from other providers. These long-term leases are called indefeasible rights of use (IRU) and telecommunications companies often buy and sell them. While technically there is nothing wrong with this practice, the manner in which the accounting is applied allows room for unethical practices.

Global Crossing stands accused by its former vice president of finance of improperly accounting for its IRU trades. The former executive alleges that the company recorded the IRUs they *sold* as revenue but recorded the IRUs they *purchased* as capital expenditures. The full amount of any revenue is recognized immediately while the cost of a capital expenditure is spread out gradually over the life of the lease. The net result is, to the layman's eye, operating results that look better than they should, and a balance sheet that has difficult-to-understand assets and liabilities. In the first six months of 2001, nearly 20 percent of Global Crossing's $3.2 billion in revenue came from such IRU deals.

The vice president was one of 1,200 employees terminated by the company in November 2001 as a cost-cutting measure. However, his attorney threatened to sue Global Crossing, claiming he was fired only because he asked questions about the company's accounting practices. The company denies that claim.

Sources: adapted from Jeremy Kahn, "How Telecom's Bad Boy Did It," *Fortune*, March 4, 2002, accessed at www.fortune.com; "FBI, SEC Investigating Global," *CNN Money*, February 8, 2002, accessed at www.cnnmoney.com; George A. Chidi, Jr., "Bankrupt Global Crossing Issues Q4 Warning," *Computerworld*, February 27, 2002, accessed at www.computerworld.com.

As you read this chapter, consider the following questions:

1. What are contingent workers and how are they frequently employed in the information technology (IT) industry?

2. What are some of the key ethical issues associated with the use of contingent workers, including H-1B and offshore outsourcing companies?

3. What is whistle-blowing and what are some of the ethical issues associated with this action?

4. What is an effective whistle-blowing process?

USE OF NON-TRADITIONAL WORKERS

Many companies today are focused on short-term over-employment problems such as corporate downsizing, and other issues prominent when the economy is sluggish. When they take a long-range view, however, they may be surprised to learn that demographers and economists think that labor shortages will return once the economy improves. What's more, these shortages are likely to persist for years, even decades, due to the huge numbers of baby boomers retiring and fewer new workers to fill the pipeline. The growth of the 25–54-year-old labor force from 1980–2000 was 38 million, but it is projected to be just 20 million from 2000–2020.[2] The prospect of long-term labor shortages could force employers to make major workplace adjustments such as higher pay raises, hiring wars, new day-care centers, flexible working hours, incentives for older workers to continue working, and expanded corporate training programs to prepare workers for new roles.

There is probably no industry where the future employment picture is as uncertain as the IT industry. The Information Technology Association of America (ITAA), a trade association representing a broad spectrum of the United States IT industry, reports that business in 2002 will need to fill 1.2 million new IT jobs, but about half, 578,000, will go unfilled. This number is much lower than the 1.6 million shortfall it says existed during the Internet boom.[3] But the ITAA estimates seem inconsistent with the evidence of swelling ranks of unemployed IT workers, with over 2.6 million workers losing their jobs in 2001.[4] The ITAA counters that the problem is that there is always a shortage of people with specific skills.

Facing a likely long-term shortage of trained and experienced workers, employers will increasingly turn to non-traditional sources to find the IT workers with the skills required to meet their needs. They will have to make ethical decisions about terms of deciding whether to recruit new and more skilled workers from these sources, or to spend the time and money to develop their current staff to meet the needs of their business. The workers affected by these decisions will demand that they be treated in a manner that they see as fair and equitable.

Contingent Workers

The **contingent workforce** includes independent contractors; individuals brought in through employment agencies; on-call or day laborers; and workers on-site whose services are provided by contract firms. According to the Bureau of Labor Statistics, the economy now includes as many as 5.6 million contingent workers. Many millions more would be included if part-time workers were added to the mix.[5]

A firm most often uses contingent workers when there are pronounced fluctuations in its staffing needs requiring technical experts on important projects, key resources in new product development, consultants in organizational restructuring, and workers in the design and installation of new information systems. Typically, these individuals join a team composed of both full-time employees and other contingent workers for the life of the project and then move on to their next assignment. Whether they work, when they work, and how much they work depends on the company's need for them. They have neither an explicit nor implicit contract for continuing employment. In sort of an information age equivalent of the migrant farm workers, they often begin planning a move to another employer as soon as they see the project at their current employer winding down.

There are two common ways for an organization to obtain contingent workers—through temporary help or employee leasing. Firms that provide temporary help recruit, train, and test their employees and assign them to clients in a wide range of job categories and skill levels. Temporary employees fill in during vacations and illnesses, meet temporary skill shortages, handle seasonal or other special workloads, and help staff special projects. However, they are never considered an official employee of the company where they are temporarily working, nor are they eligible for company benefits such as vacation, sick pay, medical insurance and so on. Such working arrangements are appealing to many people who are looking for maximum flexibility in their work schedule and a wide variety of work experiences. Temporary workers are often paid a higher hourly wage than the full-time employees doing equivalent work and who are receiving considerable additional compensation through company benefits.

Employee leasing is the placement by an employer of its existing work force onto the payroll of an employee leasing firm in an explicit co-employment relationship. A **co-employment relationship** is one where two employers have actual or potential legal rights and duties with respect to the same employee or group of employees. The goal of the hiring company is to "outsource" the human resource activities such as payroll, training, and the administration of employee benefits to the employee leasing company. Employee leasing services are subject to special regulations dealing with workers' compensation and unemployment insurance. Since the workers are technically employees of the leasing firm, they can be eligible for some company benefits through this firm.

Advantages of Using Contingent Workers

When a firm employs contingent workers, it usually does not pay for benefits (retirement, medical, and vacation benefits). In addition, the company can adjust the number of such workers to provide the level of staffing consistent with the needs of the business. Thus, the company does not need to pay the salary of a contingent worker regardless of whether or not they are busy; it simply lets them go when they are no longer needed. The company cannot do this as easily with its full-time employees without creating a great deal of ill will and lowering employee morale. Moreover, since many contingent workers already are specialists in performing a particular task, the firm does not customarily incur training costs. Therefore, the use of contingent workers enables the firm to meet its staffing needs more efficiently, lower its labor costs, and respond more quickly to changing market conditions.

Issues Raised by the Use of Contingent Workers

On the downside, the contingent worker may lack a strong relationship with the firm, which can result in low commitment to the company and its projects along with a high turnover rate. Although temporary workers don't need technical training if they are specially qualified for a temporary job, they do gain valuable practical experience working within a particular company's structure and culture, and that experience is lost when the temporary worker goes away at the project's completion.

Deciding When to Use Contingent Workers

When management decides to use contingent workers for a project, they should recognize that they are making a trade-off between completing a single project quickly and cheaply (in the short term) versus developing people in their own organization. If the project requires unique skills that are not likely to be used on future projects, there is little reason to take the additional time and incur the costs required to develop those skills in full-time employees.

If the staffing required for a particular project is truly temporary and those workers will not be needed for future work, the use of contingent workers is a good approach. In such a situation, using contingent workers avoids hiring new employees and then later firing employees when staffing needs go down.

Management should think twice about using contingent workers in situations where they are likely to learn corporate processes and strategies that are key to making the company successful. It is next to impossible to control the passing on of such information from contingent workers to their subsequent employers. This can be damaging if that next employer is a key competitor.

While using contingent workers can often be the most flexible and cheapest way to get the job done, the use of such workers can raise ethical and legal issues about the relationship between the staffing firm, its employees, and its customers, including the potential liability of the customer for the withholding of payroll taxes, payment of employee retirement benefits and health insurance premiums, and administration of workman's compensation of the staffing firm's employees. Depending on how closely workers are supervised and how the job is structured, contingent workers can be viewed as permanent employees by the Internal Revenue Service, the Labor Department, or a state's worker compensation and unemployment agencies.

In January 2001, Microsoft agreed to pay a $97 million settlement to some 10,000 so-called "permatemps" (temporary workers employed for an extended length of time). They had worked as software testers, graphic designers, editors, technical writers, receptionists, and office support staffers. Some had been temporary workers at Microsoft for several years.[6] The Vizcaino v. Microsoft class action lawsuit was filed in federal court in 1992 by eight former workers who claimed that they, and thousands more permatemps, had been illegally shut out of a stock purchase plan that allowed employees to buy Microsoft stock at a 15 percent discount. Microsoft shares had skyrocketed in value in the decade of the 1990s and split eight times. The sharp appreciation in the stock price meant that, had they been eligible, some of the temporary workers in the lawsuit could have earned more money from stock gains than they received in salary while at Microsoft.

The Vizcaino v. Microsoft lawsuit dramatically illustrates the cost of misclassifying an employee and (intentionally or unintentionally) violating laws covering workers, compensation, taxes, unemployment insurance, and overtime. The key lesson to be learned from this litigation is that even if contractors sign an agreement saying that they are contractors and not employees, the agreement is not the deciding factor. The deciding factor is how much control the company exercises over how the contractor or temporary employee performs his or her work. The degree of control factors that determine whether or not someone is an employee include:

- Does the individual have the right to control the manner and means of accomplishing the desired result?
- Over how long a period of time has the individual been employed?
- What is the amount of time the individual has worked?
- Does the individual provide his/her own tools and equipment?
- Is the individual engaged in a distinct occupation or an independently established business?
- Is the method of payment by the hour or by the job?
- What is the degree of skill required to complete the job?
- Does the individual hire employees to help?

The Microsoft ruling means that employers of contingent workers face a choice: enroll the contingent workers in company benefit plans or surrender control over them to agencies. The second alternative means that agencies must hire and fire them, promote and discipline them, do performance reviews, decide wages, and tell them what to do on a daily basis. Read the Manager's Checklist in Table 8-1 for further advice in regards to the use of contingent workers. The preferred answer to each question is "Yes."

Table 8-1   Manager's checklist for the use of contingent employees

Questions	Yes	No
Have you reviewed the definition of employee in your company's pension plan and policies to ensure that it is not so broad as to encompass contingent workers within the definition, thus entitling the worker to benefits?	——	——
Are you careful not to use contingent workers on an extended basis—do you make sure that the assignments are finite, with break periods in between?	——	——
Do you use contracts designating the worker as a contingent worker?	——	——
Are you aware that the actual circumstances of the working relationship will determine whether the worker is considered an employee in various contexts and the label assigned to a particular contingent worker may not be accepted as accurate by a government agency or a court?	——	——
Do you avoid telling contingent workers where, when, and how to do their jobs?	——	——
Do you make sure that contingent workers use their own equipment and resources such as computers and e-mail accounts?	——	——
Do you avoid providing training for your contingent workers?	——	——
When leasing employees from an agency, do you let the agency do its job—do you avoid asking to see resumes or getting involved with compensation, performance feedback, counseling, or day-to-day supervision?	——	——
If you are leasing employees, do you use a leasing company that offers its own benefits plan, deducts payroll taxes, and provides required insurance?	——	——

H-1B Workers

The H-1B visa is a temporary working visa granted by the Immigration and Naturalization Service (INS) for people who work in specialty occupations—jobs that require a four-year bachelor's degree or higher in a specific field, or the equivalent experience. Many companies turn to the use of H-1B workers to meet critical business needs. Employers hire H-1B professionals to obtain essential technical skills and knowledge not readily found in the United States. H-1B workers may also be used to fill temporary shortages of needed skills. Employers often need H-1B professionals to provide special expertise in overseas markets or topics that enable United States businesses to compete globally.

Individuals can work in H-1B status for a United States employer for a total of six years. Recent legislation allows H-1B visa holders to stay in the United States for up to seven years if they have a pending application for a green card, the employment-based permanent visa.[7] While H-1B temporary professionals comprise less than .1 percent of the United States workforce of more than 140 million people, a large percentage of them are employed as computer programmers. The top five source countries for H-1Bs currently are India, China, Canada, the United Kingdom, and the Philippines.

During 2001, United States companies hired 2.1 million IT workers, but they fired 2.6 million, reducing the overall IT workforce by about five percent, to 9.9 million workers, according to a national workforce study released by the ITAA.[8] As unemployment in the IT sector climbs to near record levels, displaced workers challenge whether the United States needs to continue bringing tens of thousands of H-1B foreign IT workers here each year. Most business managers, however, say criticizing the H-1B program conceals the real issue, which is finding qualified people, wherever they are, for increasingly challenging work. Some human resource managers and educators are concerned that the continued use of the H-1B program may be a symptom of a fundamental problem—the United States is not developing sufficient IT people with the right skills to meet the needs of corporate America.[9]

The United States government has implemented changes in their treatment of the H-1B worker since the September 11, 2001 terrorist attacks. The INS stopped providing statistics on the number of H-1B workers and will not identify the leading H-1B employers.[10] Also, the United States Defense Department has announced that it intends to ban the use of non-United States workers from working on sensitive, but unclassified technology projects.[11]

Application Process

Most ethical companies make their decision to hire someone based on how well he or she fills the qualifications for the job. In such companies, issues concerning the need for obtaining an H-1B visa are handled subsequent to the decision to hire the best available candidate. To receive an H-1B visa, the individual must have a job offer from an employer who is willing to sponsor him or her. There are two application stages: The Labor Condition Attestation (LCA) and the H-1B visa application. The company files an LCA with the Labor Department stating the job title, geographic area in which the worker is needed, and salary to be paid. The LCA is administered by the Department of Labor's Wage and Hour Division; its purpose is to ensure that the wage the individual will be paid as a foreign worker will not undercut the wage of an American worker. After the LCA is approved, the employer may then apply to the INS for the H-1B visa, identifying specifically who will fill the position and stating that person's skills in order to prove his or her qualifications for the job. A candidate cannot be hired until after the INS has processed the application, which can take from several days to several months.

Companies whose work force consists of more than 15 percent H-1B workers face further hurdles before they can hire additional H-1B workers. They must prove that they first tried to find United States workers before they can hire an H-1B. This can be done by showing copies of employment ads placed in newspapers and/or periodicals.

They also must confirm that they are not hiring the H-1B worker after having laid off a similarly situated U.S. worker. Employers must attest to the above protections by affirmatively filing with the Department of Labor (DOL) and by maintaining a file open to the public. Failure to comply with DOL regulations can result in an audit and fines in excess of $1000 per violation, payment of back wages, and ineligibility to participate in immigration programs.

The American Competitiveness in the Twenty-First Century Act contains a provision that allows current H-1B holders to start working for an employer as soon as his or her petition is filed. Therefore, if a company wishes to hire a critical individual already under H-1B status, it can do so quickly rather than wait three to five months.

Using H-1B Workers Instead of United States Workers

Heavy reliance on the use of H-1B workers lessens the incentive for United States companies to educate and develop their own work force. Some managers reason that as long as skilled foreign workers can be found to fill critical positions, why spend thousands of dollars and take months to develop their current United States workers? While such logic may appear sound for short-term hiring decisions, its implementation can come across as callous and cold-hearted, making it difficult for laid-off workers to accept. It also does nothing to develop the strong core of permanent IT workers that the United States needs as the economy expands and capital investment increases in IT products and services.

Potential "Exploitation" of H-1B Workers

Based on reports that Indian programmers working under H-1B visas typically earn up to 30 percent less than their American peers, it is apparent that some organizations are unethically exploiting foreign workers.[12] The Labor Department received 269 complaints involving abuse of H-1B workers in 2001, up from 140 the year before. It found violations in 54 cases and ordered companies to pay more than $1.3 million in back wages.[13]

Salary abuse occurs even though companies applying for H1-B visas must offer a wage that is not less than five percent below the average salary for the occupation. One reason that unethical companies can get around the "average salary requirement" is that wages in the IT field vary greatly. Determining what is an appropriate wage is an imprecise science at best. For example, an H1-B worker may be classified as an entry-level IT worker and yet be used to fill a position of an experienced worker making $10,000 - $30,000 more per year.

If at the end of the six-year visa term, the candidate is not granted a green card, the firm loses its experienced H-1B worker without having developed a permanent employee. Indeed, it is this stopgap nature of the visa program that leaves both applicants and sponsoring companies unfulfilled. The program provides the foreign IT worker no certainty of what his or her future holds. Many foreign workers find themselves unemployed and forced to uproot their families and return home, their entire life turned upside down. In some countries, there is even a stigma attached in the local society because the individual could not make it in America.

One critical aspect of training, which even highly skilled and experienced H-1B workers can require, is help using English as a second language. Often managers don't fully understand how critical it is to assist workers for whom English is a second language. This factor should be considered when weighing the use of a foreign worker versus a non-foreign worker. Without help, workers for whom English is a second language often become frustrated in meetings. They can't contribute fully in brainstorming sessions. The rapid pace of communications in most business settings is simply too quick and too full of idiomatic expressions. In addition, workers who are not fluent in English may find it difficult and uncomfortable to engage in social interaction. Even something as simple as having lunch with coworkers can be distressing. As a result, such workers often prefer to remain isolated and work alone. Even worse, they may create their own little cliques and speak their own language. All this can be bad for morale and lead to division within a project team rather than unity.

Managers of foreign-born workers should strive to improve their workers' spoken English skills and cultural understanding. The workers must be able to feel at ease and be able to interact in the social situations that are an integral part of being a member of a team. American-born workers must also be trained to be aware of and sensitive to differences in culture.

Offshore Outsourcing

Outsourcing (sometimes called managed services or facilities management) is yet another approach to meeting staffing needs. With **outsourcing**, services are provided to one company by an outside organization that has expertise in operating a specific client function. A company contracts with the firm to perform those specialized functions on an ongoing basis. Examples include contracting with an organization to operate a computer center, support a telecommunications network, or staff a computer help desk. Co-employment legal issues with outsourcing are minimal, because the company who contracts for services generally does not supervise or control the contractor's employees.

Offshore outsourcing is a variation of outsourcing where the work is done by an organization using employees who perform the work while in a foreign country. The nature of IT professionals and much of the work that they do is such that the workers can often do the work anywhere—on a company's premises or thousands of miles away in a foreign country. The ready availability of large staffs of experienced IT resources is leading to an increase in the use of offshore outsourcing within the IT industry. In addition, substantial project cost savings of up to 70 percent can be obtained largely through the lower labor costs offshore outsourcing provides. American Express, Aetna U.S. Healthcare, Compaq, General Electric, IBM, Microsoft, Motorola, Shell, Sprint, and 3M are examples of big companies that employ offshore outsourcing for functions such as help desk support, network management, and information systems development.[14]

With the high cost of United States-based application development staff and the ease with which telecommunications networks can enable communications between customer and supplier, it is now quite common to use offshore outsourcing for major programming projects. Contract programming is flourishing in Brazil, Bulgaria, China, Ireland, Israel, Malaysia, Malta, Mexico, the Philippines and Russia, among other countries. But India, with its rich talent pool, English-speaking citizenry, and low labor costs is widely acknowledged as the best source of programming skills outside of Europe and North America. India exported software to 86 countries in 1999, including the United States and many places in Europe. Its companies now employ some 260,000 software engineers.[15]

Using Offshore Outsourcing To Replace United States Employees

What would be considered low wages in the United States represents an excellent salary in other parts of the world, and some feel that companies would be foolish not to exploit this savings opportunity. Why pay a single United States IT worker a six-figure salary when you can use offshore outsourcing to hire four foreign-based workers for the same cost or even less? Another benefit of offshore outsourcing is the potential to speed up development efforts dramatically. In 1992, the state of New Mexico contracted the development of a tax system to Syntel, one of the first United States firms to successfully launch a global delivery model that enables workers to make progress on a project 24 hours per day. With technical teams working from networked facilities in different time zones, Syntel executes a virtual "24-hour workday" that saves its customers money, speeds projects to completion, and provides around-the-clock support for key software applications.[16]

Many of the same ethical issues that arise in considering the use of H-1B and contingent workers arise in the use of offshore outsourcing. Managers must make ethical decisions about whether to use offshore outsourcing firms or to spend the time and money to develop their current staff to meet the needs of their business. The workers affected by these decisions will demand that they be treated in a manner that they see as fair and equitable. In addition, like other contingent workers, offshore outsourcing employees gain valuable practical experience working within a particular company's structure and culture, and that experience is lost when the employee is reassigned at the project's completion. Lastly, offshore outsourcing does not advance the development of permanent IT workers in the United States, increasing the dependency of the United States on workers in other countries to build the IT infrastructure of the future.

The difficulty of communicating directly with people over long distances (especially those for whom English is a second language) can make it perilous to go offshore with systems work. To improve the success of an offshore outsourcing project, one must carefully evaluate the outsourcing company to ensure that its employees meet these five basic prerequisites: 1) they have expertise in the technologies involved in the project, 2) they are proficient in the employer company's native spoken language, 3) a large staff is available, 4) they have a good telecommunications setup, and 5) good on-site managers are available from the outsourcing partner. Successful projects require day-to-day interaction between software development and business teams. It is essential that the hiring company not take a hands-off approach to project management. You cannot outsource responsibility and accountability.

However, offshore outsourcing doesn't always pay off. A software development company in Cambridge, MA., went to India in search of cheap labor but instead found lots of problems. Customs officials charged huge tariffs when the company tried to

197

ship the necessary development software and manuals, and the programmers weren't nearly as experienced as they claimed to be. The code that was produced was inadequate. The company had to send a representative to India for months to work with the Indian programmers to correct the problems.[17]

Workers, especially those who have been laid off, can become very upset about their jobs being outsourced, especially to low-wage workers residing outside the United States. Members of a department being outsourced who suddenly have to transition their work to foreign nationals can become upset and non-productive. Morale among any remaining members of the department may be extremely low.

Cultural differences can also cause misunderstandings among project members. Indian programmers, for instance, are known for keeping quiet, even when they notice problems. And subtle differences in gestures can cause confusion. In the United States, shaking your head from side to side means "no." In the south of India, it means "yes." Table 8-2 provides several tips to ensure a successful offshore outsourcing experience.

Table 8-2   How to assure a successful offshore outsourcing experience
Set clear, firm business specifications for the work to be done.
Assess the probability of political upheavals or factors that might interfere with transnational information flow and ensure that they are acceptable.
Assess the basic stability and economic soundness of the outsourcing vendor and what might occur should the vendor encounter a severe financial downturn.
Establish reliable satellite or broadband communications between your site and the outsourcer's location.
Implement a formal version-control process, coordinated through a quality assurance person.
Develop and use a dictionary of terms so that there is common understanding of technical jargon.
Require that the vendor supply a project manager at the client site to overcome cultural barriers and facilitate communication with offshore programmers.
Require a network manager at the vendor site to coordinate the logistics of using several communications providers across the continents.
Obtain advance agreement on the structure and content of documentation to avoid manuals that explain how the system was built, not how to maintain it.

WHISTLE-BLOWING

Like the subject of the use of contingent workers, the subject of whistle-blowing is a significant topic in any discussion of ethics in IT. Not only do both subjects raise ethical issues, but each also has social and economic implications. How these issues are addressed can have a major and long-lasting impact on not only the individuals and employers involved, but also the entire IT industry.

Whistle-blowing is an employee's effort to attract the attention of others outside of the employee's company to a negligent, illegal, unethical, abusive, or dangerous act by the company that threatens the public interest. In some cases, corporate whistle-blowers are acting as no more than informers on their company, revealing information to enrich themselves or to gain revenge for some perceived wrong. In most cases, however, whistle-blowers are acting ethically out of a conviction to correct a major wrongdoing, often at great personal risk. The whistle-blower often has special information about what is happening based on his or her expertise or position of employment within the offending organization. Action taken by a whistle-blower can be disastrous to the person's own career and will likely affect the lives of his or her friends and family. In the extreme, whistle-blowers must choose between protecting society or remaining loyal to their employers.

Protection for whistle-blowers

Whistle-blower protection laws allow employees to alert the proper authorities to employer actions that are unethical, illegal, unsafe, or that violate specific public policies without concern for employer retribution. Unfortunately, there is no comprehensive federal law that protects all whistle-blowing employees. Instead, there are numerous laws, each one protecting a certain class of employees who engage in certain specific whistle-blowing acts in the various protected industries. To make things even more complicated, each law has different filing provisions, a different statute of limitations, and different administrative and judicial remedies. Thus, the first step in reviewing a whistle-blower's claim of retaliation is for an experienced attorney to analyze the various federal and state laws to determine if and how the employee is protected and exactly what procedures should be followed in filing a claim. A short statute of limitations is one major weakness in many statutory whistle-blower protection laws. Indeed, failure to comply with the statute of limitations is one of the favorite defenses in whistle-blower cases. For instance, the statute of limitations for reporting of some kinds of fraud is five years. The statute of limitations is generally held to start at the time an employee first learns that he or she will be retaliated against rather than the last day of employment.

The False Claims Act is a federal law that provides strong protection for whistle-blowers. See the Legal Overview for more information about this act.

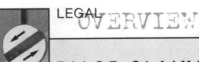 LEGAL OVERVIEW

FALSE CLAIMS ACT

The False Claims Act, also known as the "Lincoln Law," was enacted during the Civil War to combat the fraud perpetrated by companies that sold supplies to the Union Army. War profiteers were shipping boxes of sawdust instead of guns, for instance, and swindling the Union Army into purchasing the same cavalry horses several times. At the time it was enacted, the act's goal was to entice whistle-blowers to come forward by offering them a share of the money recovered.

(continued)

199

The **qui tam** ("who sues on behalf of the king as well as for himself") is a provision of the Federal Civil False Claims Act that allows a private citizen to file a suit in the name of the United States government, charging fraud by government contractors and other entities who receive or use government funds, and to share in any money recovered. In qui tam actions, the government has the right to intervene and join the legal action. If the government declines, the private plaintiff may proceed on his or her own. Some states have passed similar laws concerning fraud in state government contracts.

There are several types of cases filed as qui tam actions, including mischarging (such as submitting claims to the government for medical services provided by an attending physician when, in actuality, the service was performed by a nurse and should have been billed at a lower rate), product and service substitution, false certification of entitlement for benefits, and false negotiation (such as the submission of false costs to the government during the process of contract negotiation to justify an inflated contract price). Mischarging is the most common form of qui tam case filed. A common mischarging scenario is employee labor charged to a government contract not worked on (for example, an IT contractor charging hundreds of hours of programming effort not performed).

Violators of the False Claims Act are liable for three times the dollar amount that the government is defrauded. They can also receive civil penalties of $5,000 to $10,000 for each instance of a false claim. A qui tam plaintiff can receive between 15 and 30 percent of the total recovery from the defendant, depending on how helpful the individual was to the success of the case. Successful recoveries for a qui tam case have been as high as $150 million. Over the years, private citizen whistle-blower lawsuits have helped the government recover more than $3.5 billion from companies that allegedly defrauded the taxpayers. For example, in December 2000, The Health Company (HCA), the nation's largest for-profit hospital chain, agreed to pay $840 million in fines and penalties to resolve allegedly unlawful billing of federally and state funded health care programs. The allegations concerned billing for services provided to ineligible patients, improper billing for certain lab tests, and billing for home health services that were unnecessary or never provided.

The False Claims Act provides strong whistle-blower protection. Briefly, any employee who is discharged, demoted, harassed, or otherwise discriminated against because of lawful acts by the employee in furtherance of an action under the False Claims Act is entitled to all relief necessary to make the employee whole. In other words, such relief may include job reinstatement, double back pay, and compensation for any special damages including litigation costs and reasonable attorney's fees. Because of the high potential for significant financial recovery, one does not usually find it difficult to obtain the services of an attorney for the filing of such action. However, the provisions of the False Claim Act are quite complicated. It is extremely unwise to pursue such a claim without counsel.

Sources: adapted from "HCA Agrees to Pay $840 Million In Government's Largest Fraud Settlement Yet," Press Release, September 1, 2000, found at Taxpayers Against Fraud Web site, accessed at www.taf.org. and "Court Narrows Whistle-Blower Law," *USA Today*, May 22, 2000, accessed at www.usatoday.com.

In addition to the False Claims Act, there are other laws that protect whistle-blowers, some aimed at specific industries.

Environmental Protection

There are several major federal laws intended to protect the environment—The Clean Air Act (first passed in 1955 and major revisions in 1963, 1970, and 1990); The Toxic Substances Control Act (1976); The Clean Water Act (first passed in 1970 and amended in 1977 and 1990); The Safe Drinking Water Act (1974), and The Comprehensive Environmental Response, Compensation, and Liability Act (established the Superfund, 1980, amended in 1986 and 1994). Special provisions were added to each of these laws to protect employee whistle-blowers (both in government and private industry) so as to encourage reporting of any wrongdoing that would damage the environment.

Nuclear Workers

The Energy Reorganization Act safeguards workers in the nuclear power and nuclear weapons industries. Whistle-blowers in these industries are provided protection because of the potential for serious harm to those who work with or live near facilities producing, handling, or using nuclear power or nuclear weapons. Almost any employee in these industries, whether employed in the private sector or by local, state, or federal government is protected against actions by his or her employer aimed at discouraging the reporting of any safety problem, environmental violation, or other illegal activity. In addition, any employee who believes he or she has been discriminated against in retaliation for "blowing the whistle," is also protected under the law.

Private Sector Workers

Under traditional state common law, in the absence of an employment contract, an employee can be terminated for any reason or no reason. However, a majority of the states have created laws that protect workers from being fired because of an employee's participation in so-called "protected" activities. One example of a protected activity is the filing of a qui tam lawsuit under the provisions of the False Claims Act, charging fraud by government contractors or others who receive or use government funds. States that have these laws recognize that there is a benefit to the public for the successful prosecution of such cases and, therefore, elect to protect the whistle-blower. Employees in these states may file claims against their employers for retaliatory termination for participation in protected activities and are entitled to jury trials. If successful, they can receive punitive damage awards.

Dealing with a Whistle-Blowing Situation

Each potential whistle-blowing situation involves different circumstances, issues, and personalities. Two people in the same company facing the same situation may have different values and concerns, causing them to react somewhat differently, without either person being unethical. Therefore, it is impossible to outline an exact step-by-step procedure of how one ought to behave in dealing with a whistle-blowing situation. This section will provide a general sequence of events and highlight the key issues that should be considered by the potential whistle-blower.

1. **Assess the seriousness of the situation.** The first step is to determine the seriousness of the problem that has been observed. For someone to consider whistle-blowing, that person should have specific knowledge of a situation where someone is taking a course of action that is unethical and that represents a serious threat to the public interest. Such a threat might include loss of life or the stealing of billions of dollars. The employee should carefully and informally seek out selected and trusted resources inside the company and ask for their assessment. Do they see the situation as serious as well? Their input may help the employee see the situation from a different perspective and alleviate the concerns. On the other hand, this research may reinforce the initial suspicions, forcing the employee to face a series of difficult, ethical decisions.

2. **Attempt to address the situation internally.** Ideally the employee can get the problem out into the open and deal with it from inside the organization, creating as few bad feelings as possible. The focus should be on disclosure of the facts of the situation and how it affects others. The employee's goal should be to get the problem fixed, not place blame. Given the strong negative impact of whistle-blowing on the employee's future life, this step should not be dismissed or taken lightly. Fortunately, many problems are solved at this level without cause to even consider further, more drastic actions by the employee.

3. **Begin documentation.** Any attempt to address the situation should be documented with a written summary of the problem provided to the appropriate managers, including a statement that they either responded or clearly chose not to respond. The employee should start to keep a record of all events, facts, and insights about the situation. This record will be helpful in constructing a chronology of events if legal testimony is required in the future. Memos, correspondence, manuals, and other documents that support the employee's position should be identified and copied *before* the next step is taken. Actions taken afterwards may be too late as records may disappear and not be accessible. Documentation should be maintained and kept up-to-date throughout the whistle-blowing process.

4. **Consider escalation within the company.** The employee may be unsuccessful in his or her initial attempt to deal with the situation internally. At this point, the employee may rationalize that he or she has done all that is required by raising the issue. Some people who elect to take no further action continue to wrestle with their consciences; they can develop ulcers, drug or alcohol problems, and/or lose their peace of mind. Others may feel so strongly about the situation that they must take further action. This, is often the case when a very determined and conscientious individual faces a truly serious problem. Thus, the employee who is still concerned is forced to choose between escalating the problem and going over his or her manager's head or going outside the organization to deal with the problem. At this point, the employee may become a whistle-blower if he or she feels compelled to sound the alarm on the company because there appears to be no chance to solve the problem internally.

Going over an immediate manager's head can certainly put one's career in jeopardy. Employers may retaliate against a challenge to their management. Perhaps the organization has a strong, effective corporate ethics officer who can be trusted to give the employee a fair and objective hearing. Another choice is a senior manager who has some responsibility for the area of concern and has a reputation for being fair. However, if this senior manager does not agree with the assessment of the situation, there may be retaliation against the employee. In all but the most enlightened of work environments, the challenger is likely to be fired, demoted, or reassigned to a less-desirable position or job location. Such actions send a loud signal throughout the organization that loyalty is highly valued and that challengers will be dealt with harshly. Whether reprisal actions are ethical depends in large part on the legitimacy of the employee's issue. If the employee attempted to treat an issue much more seriously than was warranted, then the employee does deserve some sort of reprisal for exercising poor judgment and accusing management of unethical behavior.

If the senior manager refuses to deal with the problem, the options are to drop the matter or to go outside the organization to remedy the situation. Even if the senior manager agrees with the employee's position and overrules his or her boss, the employee may want to request a transfer to avoid working for the same boss, the safest thing for his or her career.

5. **Assess implications of becoming a whistle-blower.** If the employee feels he or she has made a strong attempt to resolve the problem internally and has achieved no results, the employee is now at the threshold of blowing the whistle and must stop and fully assess if he or she is prepared to go forward. Depending on the nature of the situation, significant legal fees may be expected in airing or bringing charges against an agency or company that has unlimited time and legal resources, and a lot more money. At this point, the employee has learned that the organization does not agree with the employee's call for change. Should the

employee choose to proceed, it's possible he or she will be accused of having a grievance with the employer, or of trying to profit from the accusations made. The employee may be fired; family, friends, and coworkers may turn against him or her.

There are many ethical questions that a potential whistle-blower must attempt to answer:

1. Given the high price that whistle-blowers must pay, does he or she really want to blow the whistle?
2. Has the employee exhausted all means of dealing with this problem and is whistle-blowing all that is left?
3. Is he or she violating the obligation that employees owe employers to be loyal and work for the best interest of the company, not against it?
4. Will the public exposure of corruption and mismanagement in the organization really correct the underlying cause of these problems and protect others from harm?

From the moment an employee becomes known as a whistle-blower, everything about that person becomes fair game and a very public battle may ensue. Attacks on personal integrity and character can be expected, and possibly may even be publicized in the news. Friends and family will hear these accusations and may become upset or alarmed. Thus, before blowing the whistle, family and friends should be consulted for advice and to test their willingness to be supportive. They should be told what to expect to prevent them from being surprised by the employee's or the employer's future actions.

Support people or groups such as elected officials and professional organizations should also be consulted. For example, the National Whistle-blower Center can provide referrals for counsel to whistle-blowers nation-wide and provides education about the rights of whistle-blowers.

6. **Use experienced resources to develop an action plan.** A whistle-blower should consult with competent, qualified, legal counsel with experience in whistle-blowing cases. They will determine what statutes and laws apply to the case. There may be more than one provision that may apply depending on the agency, employer, state involved, and the nature of the case. Legal counsel should determine the statute of limitations for reporting the offense as well as the length and nature of whistle-blower protection. Before blowing the whistle, the employee should get an honest assessment of the soundness of his or her legal position. Fees and other costs should also be estimated if the plan is to file a lawsuit.

7. **Execute the action plan.** If the whistle-blower chooses to continue with the matter legally, the claim should be pursued based on the research and decisions previously made with legal counsel. The consequences of every action should be understood. Sometimes, if the whistle-blower wants to remain unknown, the safest course of action is to leak information about the situation anonymously to the press. The problem with

this approach is that people often do not take such anonymous claims seriously. In most cases, working directly with the appropriate regulatory agencies and legal authorities is more likely to result in action. These actions include the imposing of fines, halting of operations, or other proceedings designed to draw the offending organization's immediate attention.

8. **Live with the consequences.** Whistle-blowers must be on guard against some form of retaliation. Coworkers may be asked to help discredit the person, and he or she may even be threatened. The whistle-blower must safeguard against possibly being "set up" for a situation that would be damaging. For example, management may attempt to have the whistle-blower transferred, demoted, or fired for breaking some minor rule, perhaps arriving late to work or leaving early. They may argue that such behavior has been ongoing to justify their actions. Hopefully, the whistle-blower and his or her attorney will have planned a good strategy to counteract such actions and there will be some recourse that can be taken under the law.

Summary

1. What are contingent workers and how are they frequently employed in the IT industry?

    The contingent workforce includes independent contractors; individuals brought in through employment agencies; on-call or day laborers; and workers on-site whose services are provided by contract firms. Employers hire contingent workers to obtain essential technical skills or knowledge that is relatively unique and not readily found in the United States or to meet temporary shortages of needed skills. Some people contend that employers exploit contingent workers, especially H-1B foreign workers, to obtain skilled workers at lower than the going cost.

2. What are some of the key ethical issues associated with the use of contingent workers, including H-1B and offshore outsourcing companies?

    The use of contingent workers enables the firm to meet its staffing needs more efficiently, lower its labor costs, and respond more quickly to changing market conditions. Facing a likely long-term shortage of trained and experienced workers, employers will increasingly turn to non-traditional sources to find the IT workers with the skills required to meet their needs. They will need to make ethical decisions in terms of deciding whether to recruit new and more skilled workers from these sources or to spend the time and money to develop their current staff to meet the needs of their business. The workers affected by these decisions will demand that they be treated in a manner that they see as fair and equitable.

    The use of contingent workers can raise ethical and legal issues about the relationship between the staffing firm, its employees, and its customers, including the potential liability of the customer for the withholding of payroll taxes, payment of employee retirement benefits and health insurance premiums, and administration of workman's compensation of the staffing firm's employees.

3. What is whistle-blowing and what are some of the ethical issues associated with this action?

    Whistle-blowing is an employee's effort to attract the attention of others outside of the employee's company to a negligent, illegal, unethical, abusive, or dangerous act by the company that threatens the public interest.

    There are many ethical implications that a potential whistle-blower must consider: whether the high price of whistle-blowing is worth it; whether all means of dealing with the problem have been exhausted before resorting to whistle-blowing; whether whistle-blowing violates the obligation of loyalty that employees owe to employers; and whether public exposure of the problem will actually correct the underlying cause of the problem and protect others from harm.

4. What is an effective whistle-blowing process?

    There is an eight-stage process for effective whistle-blowing: 1) assess if this is a potential whistle-blowing incident, 2) attempt to address the situation internally, 3) begin documentation, 4) consider escalating the situation within the company, 5) assess the implications of becoming a whistle-blower, 6) use experienced resources to develop an action plan, 7) execute your action plan, 8) live with the consequences.

Review Questions

1. What is the contingent workforce and in what kind of roles are these workers employed most effectively?
2. What are the key factors that determine whether or not someone is an employee?
3. Briefly discuss the advantages and disadvantages of using a contingent workforce. Discuss the advantages of being a contingent worker.
4. Briefly summarize the legal areas in which co-employment issues can arise.
5. What are some of the ethical issues associated with the use of contingent workers?
6. What is an H-1B visa? What are the requirements for a worker to qualify for an H-1B visa?
7. What is the application process for an H-1B visa? Why is it divided into two phases?
8. Briefly summarize the primary concerns and ethical issues surrounding the use of H-1B workers.
9. What is offshore outsourcing? What are some of the advantages of taking this approach in the area of application development?
10. What ethical issues can the use of offshore outsourcing raise?
11. What is a global delivery model? Why is such a model attractive to many companies?
12. What is whistle-blowing? Under what conditions is it justified?
13. What legal protection is available to whistle-blowers? What is a frequent weakness in these statutes?
14. Briefly summarize the stages in the whistle-blowing process.

Discussion Questions

1. What advice would you offer a manager within your company who is considering leasing employees to meet a critical business need?
2. Do you think the use of H-1B technical workers is justified in a company that is laying off many of its United States-based technical workers?
3. What factors would you weigh in deciding whether or not to employ offshore outsourcing on a particular project?
4. Which step(s) of the whistle-blowing process are the most important? Why?

What Would You Do?

1. Dr. Jeffrey Wigand is a whistle-blower who was fired from his position of vice president for Research and Product Development at Brown & Williamson Tobacco Company in 1993. He was interviewed for a segment of the CBS show "60 Minutes" in August 1995, but, in a highly controversial decision, the interview was initially not aired. CBS management wouldn't broadcast the interview because they were worried about the possibility of a multi-billion dollar lawsuit for tortious interference, that is, interfering with Wigand's confidentiality agreement with Brown & Williamson. The interview was finally aired on February 4, 1996 after the *Wall Street Journal* published a confidential November 1995 deposition Wigand gave in a Mississippi case against the tobacco industry that repeated many of the charges he made to CBS. In the interview, Wigand said that Brown & Williamson had scrapped plans to make a safer cigarette and continued to use a flavoring in pipe tobacco that was known to cause cancer in laboratory animals. Wigand also charged that tobacco industry executives testified untruthfully before Congress about

tobacco product safety. Wigand suffered greatly for his actions. He lost his job, his home, his family, and his friends. Visit his Web site at www.jeffreywigand.com and answer these questions. (The "Insider" is a 1999 R-rated movie based on Wigand's experience. You may also wish to view this movie.)

a. What motivated Wigand to take a position as head of Research and Product Development for a tobacco company and then five years later denounce the industry's efforts to minimize the health and safety issue of tobacco use?

b. What whistle-blower actions did Dr. Wigand take?

c. If you were in Dr. Wigand's position, what would you have done?

2. Microsoft is a major user of temporary workers and, to minimize legal issues, sought to ensure that those workers were not mistaken about their place within the company. The temporary agencies that Microsoft used provided the workers with handbooks that laid out the ground rules in explicit detail. Temporary workers were barred from using company-owned athletic fields—for insurance reasons, the company explained. At some Microsoft facilities, temporaries were told that they could not drive their cars because it would create parking problems for regular workers. Instead, they were told to take the bus. In addition, they were told not to participate in the social clubs such as chess, tai chi or rock-climbing that were open only to regular employees. They also were not permitted to buy goods at the company store or to attend parties given for regular employees, such as the private screening of the latest "Star Wars" film, or company-wide meetings held at the Kingdome stadium in Seattle. In addition, their e-mail addresses were to be preceded with "a-" before their names so that anyone with whom they communicated would know their non-permanent status in the company.

Imagine that you are a senior manager in Human Resources at Microsoft and that you have been asked to respond to a group of temporary workers complaining about working conditions. How would you handle this?

3. Catalytic Software, a United States-based IT outsourcing firm with offices both in Redmond, WA and Hyderabad, India, is looking at tapping India's large supply of engineers as contract software developers for IT projects. However, instead of just outsourcing projects to local Indian software development companies, as is the common practice followed by United States companies, Catalytic is developing a completely self-contained company community near Hyderabad. Spread over 500 acres, the town is called New Oroville, and is a self-sustaining residential and office community that is expected to house around 4,000 software developers and their families, as well as 300 support personnel for sanitation, police, and fire.

The goal of this high-tech city is to knock down the barriers to success that large-scale technology business encounters in India. By building a company community, Catalytic ensures that it will have enough qualified employees to staff round-the-clock shifts. The company expects this facility to attract and keep top professionals from all over the world. Building a company town also solves the problem of transportation of a 4,000-strong staff. Because of the terrible roads, the commute from Hyderabad, 25 kilometers from New Oroville, would take roughly an hour.

Catalytic will provide private homes with private gardens, all within a short walk of work, school, recreation, shopping and public facilities. Each house will include cable TV, telephones, and a fiber-optic data pipeline connecting to the Internet so that employees can work efficiently even when they are at home. (Employees will be awarded bonuses for working overtime.) New Oroville will have four indoor recreational complexes and six large retail complexes. There will be ample green space, including one central park and four larger perimeter parks for outdoor exercise and recreation.

You have just completed a job interview with Catalytic Software for a position as Project Manager. They have offered you a 25 percent increase over what you are currently making. Your position will be based in Redmond and will involve providing United States-based project management for customer projects. A requirement of the position is that you spend the first year with Catalytic in the New Oroville facility to learn their methods, culture, and people. (You may take your entire family, if you wish, or have two, three-week company-paid trips back to Redmond.) Why do you think the temporary assignment in New Oroville is a requirement? What else do you need to know in considering this position? Would you accept this position? Why or why not?

4. The Recording Industry Association of America (RIAA) is tracking down companies whose employees create digital jukeboxes by downloading tunes onto company file servers and sharing them with coworkers. For example, in April 2002, RIAA announced a $1 million settlement with one company, even though the company had immediately removed its employees' digital jukebox after being contacted by the RIAA. The RIAA insisted that the company either pay up or go to court, where the penalty might have been higher.[18] Similarly, the Business Software Alliance (BSA) is a watchdog group that represents most major software manufacturers in combating the illegal copying of software. Both the RIAA and BSA get most of their tips from disgruntled ex-employees who call its anonymous hot line. If you were an employee of a company that was running afoul of the RIAA or BSA, would you call either one to report your company or would you attempt to work out the issue internally? What if there were a $10,000 reward? What if the company had recently let you go due to a need to reduce company expenses?

Cases

1. Software to Assist in Managing Contingent Workers

Many organizations are turning to Recruitment and Applicant Tracking Systems (RATS), software packages that can be used to manage and control the dollars spent on contingent workers. These systems are capable of tracking which workers and which suppliers provide quality service, how much each department is spending on contingent workers, and how many contingent workers are on the job. This information improves hiring decisions by enabling managers to better gauge consulting and service costs and quality to choose the best providers.

Most RATS centralize the hiring process, keeping track of which suppliers are providing the employees, and determining where the hiring company can get the most leverage in terms of cost reduction and services. Many of the packages offer the added benefit of streamlining the approval of job requisitions with automated workflow. Some even have built-in checks and balances that prevent hiring managers from circumventing the company's hiring rules. A common violation is that the hiring manager of one department contracts with someone to work on a specific project because he wants to get him a job or he is a friend of a friend. Most of the packages also offer back-end integration to financial and human resources systems to streamline payment processes.

Some RATS allow hiring managers to rate the contingent workers, which not only provides feedback to suppliers but also lets other hiring managers within the same company determine whether a contingent employee should be rehired in another area of the firm. Managers can then point-and-click on the jobs they need filled and the agencies they prefer to use. If the system keeps a history of all workers, the manager can even request a specific worker who performed well on a previous assignment, or find someone the company used three months ago. The time it takes to process an order for a contingent worker can take less than a minute.

A future enhancement for these systems is to include benefits administration, training, and validation of the skill sets potential employees purport to have. Doing so will further streamline the hiring process.

Even though most companies can achieve significant savings and other benefits, it is often difficult to get human resources, hiring managers, and other company employees to use RATS because a significant culture change is required. There is frequently resistance to the change in the vendor-management concept, with hiring managers asking why they can't use a certain consulting service anymore. Suppliers also resist the change to a more automated system, fearing that they will lose control and potential income.

Questions:

1. What are the key advantages associated with the use of a RATS?
2. Does the use of a RATS increase or decrease the chances that a company will operate in a more ethical manner in dealing with contingent workers? Why do you think so?
3. What additional features do you think should be added to the RATS application to increase its effectiveness? Can you think of any specific features to add to help managers better address the concerns of full-time workers toward the use of contingent workers?

Sources: adapted from Karen D. Schwartz, "Managing Human Capital," *Computerworld*, April 16, 2001, accessed at www.computerworld.com; "Automating Internal Project Management and Staffing," *Computerworld*, April 16, 2001; accessed at www.computerworld.com; Katherine Jones and David Alschufer, Aberdeen Group, "Hourly Hiring Management Systems: Improving the Bottom Line for Hourly Worker-Centric Enterprises," Aberdeen White Paper, June 2002, accessed at www.aberdeen.com; Elisabeth Goodridge, "Workforce Management Draws VC Dollars," *InformationWeek*, October 29, 2001, accessed at www.informationweek.com.

2. Sherron Watkins, Enron "Whistle-Blower"

Sherron Watkins was the vice president of corporate development at Enron. While analyzing sales of Enron assets, she was alarmed to find questionable accounting mechanisms being used to hide company debt. She also thought it was inappropriate for the company to use its stock to affect its income statement, as those mechanisms did by using company stock shares to hedge sales.

In March 2000, one of Watkins' friends at Enron, Jeff McMahon, had complained to CEO Jeffrey Skilling about the accounting irregularities controlled by Watkins' boss, Chief Financial Officer Andrew Fastow. But instead of addressing McMahon's complaints, Skilling transferred him to a different job at Enron. Based on her friend's experience and the fact that Fastow had a close relationship with Skilling, Watkins felt there was nothing she could do to stop her boss.

Things came to a head in August 2001. Watkins had learned that a large portion of Enron's profits relied on its own stock price. By August, Enron stock had dropped to $38/share, down from nearly $80, portending future trouble. So when Skilling resigned that month, Watkins wrote an anonymous, but detailed seven-page memo to Enron Chairman Ken Lay warning that the company could "implode in a wave of accounting scandals." She also warned that there were potential whistle-blowers within the company who might go to authorities.

After Watkins identified herself as the author of the memo, Lay met with her to discuss her concerns. Lay then referred the matter to Enron auditor Arthur Andersen and the law firm Vinson & Elkins. Both concluded that the transactions Watkins questioned were proper, although they might look bad. Watkins also testified that after her meeting with Lay, her boss, Andrew Fastow, wanted her fired and her computer seized. Instead, she was transferred from his group into a dead-end Enron job.

Watkins' testimony actually helped provide cover for Lay and the Enron board. The fact that Watkins "warned" Lay presumes that he knew nothing and needed to be warned. This presumption could prove to be a key element in Lay's legal defense. Watkins also said that she told Lay that Skilling, Fastow, and other executives had fooled him and the board.

Enron filed for Chapter 11 bankruptcy on December 2, 2001. Watkins' memo was not made public until congressional investigators released it to the public six weeks later, some six months after she initially wrote it. Some claim that she may have written the memo to protect herself.

Questions:

1. Would you classify Sherron Watkins as a whistle-blower? Do you think that she acted ethically? Why or why not?

2. Identify several different reasons why Watkins might have been motivated to finally come forward. Which of these do you suspect was the strongest motivator? Why?

3. With the benefit of 20-20 hindsight, what might an Enron employee have done to prevent the company's collapse? Would it be necessary for the employee to be positioned high in the Enron hierarchy? When would they have had to take this action? Why do you think no one did this?

Sources: adapted from Frank Pellegrini, "Person of the Week: 'Enron Whistleblower' Sherron Watkins," *Time,* January 18, 2002, accessed at www.time.com; Greg Farrell, "Watkins to Testify About Memo Thursday," *USA Today*, February 14, 2002, accessed at www.usatoday.com; and Dan Ackman, "Sherron Watkins Had Whistle, But Blew It," *Forbes*, February 14, 2002 accessed at www.forbes.com.

Endnotes

1 Aaron Berstein, "Too Many Workers? Not For Long," *BusinessWeek*, May 20, 2002, pp. 126–130.

2 Aaron Berstein, "Too Many Workers? Not For Long," *BusinessWeek*, May 20, 2002, pp. 126–130.

3 Diane Rezendes Khirallah, "IT Jobs Don't Come Easy," *InformationWeek*, May 13, 2002, accessed at www.informationweek.com.

4 Julia King, "Study: IT Workforce Down 5%," *Computerworld*, May 2, 2002, accessed at www.computerworld.com.

5 Patricia Mendels, "Allowing Temps to Organize," *BusinessWeek Online*, September 14, 2000, accessed at www.businessweek.com.

6 Loretta W. Prencipe, "Review Temporary Worker's Status," *InfoWorld*, January 8, 2001, accessed at www.infoworld.com.

7 Patrick Thibodeau, "House, Senate Vote to Increase H-1B Visa Cap," *Computerworld*, October 9, 2000, p. 7.

8 Julia King, "Study: IT Workforce Down 5%," *Computerworld*, accessed at May 2, 2002, www.computerworld.com.

9 Diane Rezendes Khirallah, "Where Does H-1B Fit?" *InformationWeek*, February 4, 2001, pp. 34–42.

10 Diane Rezendes Khirallah, "Guess Who's Using H-1B Visas," *InformationWeek*, February 11, 2002 accessed at www.informationweek.com.

11 Elisabeth Goodridge, "Foreign IT Workers Banned at Defense Department," *InformationWeek*, March 7, 2002, accessed at www.informationweek.com.

12 Diane Rezendes Khirallah, "President to Sign H-1B Visa Bill," *InformationWeek*, October 9, 2000, p. 251.

13 Diane Rezendes Khirallah, "Where Does H-1B Fit?" *InformationWeek*, February 4, 2001, pp. 34–42.

14 Drew Robb, "Offshore Outsourcing Nears Critical Mass," *InformationWeek*, June 12, 2000, pp. 89–98.

15 Clive Couldwell, "UK Firms See Business Case in Outsourcing IT to India," *IT Week*, March 7, 2000, accessed at www.zdnet.co.uk/itweek.

16 Lynda Radosevich, "Offshore Development Shipping Out," *IT Week*, September 1, 1996, accessed at www.zdnet.co.uk/itweek.

17 Lynda Radosevich, "Offshore Development Shipping Out," *IT Week*, September 1, 1996, accessed at www.zdnet.co.uk/itweek.

18 Jeremy Kahn, "After Napster," *Fortune*, June 24, 2002, accessed at www.fortune.com.

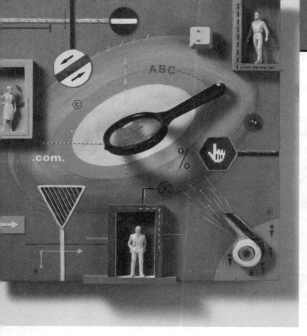

Appendix A

Association for Computing Machinery (ACM) Code of Ethics and Professional Conduct

Adopted by ACM Council 10/16/92.

Preamble

Commitment to ethical professional conduct is expected of every member (voting members, associate members, and student members) of the Association for Computing Machinery (ACM).

This Code, consisting of 24 imperatives formulated as statements of personal responsibility, identifies the elements of such a commitment. It contains many, but not all, issues professionals are likely to face. Section 1 outlines fundamental ethical considerations, while Section 2 addresses additional, more specific considerations of professional conduct. Statements in Section 3 pertain more specifically to individuals who have a leadership role, whether in the workplace or in a volunteer capacity such as with organizations like ACM. Principles involving compliance with this Code are given in Section 4.

The Code shall be supplemented by a set of Guidelines, which provide explanation to assist members in dealing with the various issues contained in the Code. It is expected that the Guidelines will be changed more frequently than the Code.

The Code and its supplemented Guidelines are intended to serve as a basis for ethical decision making in the conduct of professional work. Secondarily, they may serve as a basis for judging the merit of a formal complaint pertaining to violation of professional ethical standards.

It should be noted that although computing is not mentioned in the imperatives of Section 1, the Code is concerned with how these fundamental imperatives apply to one's conduct as a computing professional. These imperatives are expressed in a general form to emphasize that ethical principles which apply to computer ethics are derived from more general ethical principles.

It is understood that some words and phrases in a code of ethics are subject to varying interpretations, and that any ethical principle may conflict with other ethical principles in specific situations. Questions related to ethical conflicts can best be answered by thoughtful consideration of fundamental principles, rather than reliance on detailed regulations.

1. GENERAL MORAL IMPERATIVES

As an ACM Member I will ....

1.1 Contribute to Society and Human Well-Being

This principle concerning the quality of life of all people affirms an obligation to protect fundamental human rights and to respect the diversity of all cultures. An essential aim of computing professionals is to minimize negative consequences of computing systems, including threats to health and safety. When designing or implementing systems, computing professionals must attempt to ensure that the products of their efforts will be used in socially responsible ways, will meet social needs, and will avoid harmful effects to health and welfare.

In addition to a safe social environment, human well-being includes a safe natural environment. Therefore, computing professionals who design and develop systems must be alert to, and make others aware of, any potential damage to the local or global environment.

1.2 Avoid Harm to Others

"Harm" means injury or negative consequences, such as undesirable loss of information, loss of property, property damage, or unwanted environmental impacts. This principle prohibits use of computing technology in ways that result in harm to any of the following: users, the general public, employees, and employers. Harmful actions include intentional destruction or modification of files and programs leading to serious loss of resources or unnecessary expenditure of human resources such as the time and effort required to purge systems of "computer viruses."

Well-intended actions, including those that accomplish assigned duties, may lead to harm unexpectedly. In such an event the responsible person or persons are obligated to undo or mitigate the negative consequences as much as possible. One way to avoid unintentional harm is to carefully consider potential impacts on all those affected by decisions made during design and implementation.

To minimize the possibility of indirectly harming others, computing professionals must minimize malfunctions by following generally accepted standards for system design and testing. Furthermore, it is often necessary to assess the social consequences of systems to project the likelihood of any serious harm to others. If system features are misrepresented to users, coworkers, or supervisors, the individual computing professional is responsible for any resulting injury.

In the work environment the computing professional has the additional obligation to report any signs of system dangers that might result in serious personal or social damage. If one's superiors do not act to curtail or mitigate such dangers, it may be necessary to "blow the whistle" to help correct the problem or reduce the risk. However, capricious or misguided reporting of violations can, itself, be harmful.

Before reporting violations, all relevant aspects of the incident must be thoroughly assessed. In particular, the assessment of risk and responsibility must be credible. It is suggested that advice be sought from other computing professionals. See principle 2.5 regarding thorough evaluations.

1.3 Be Honest and Trustworthy

Honesty is an essential component of trust. Without trust an organization cannot function effectively. The honest computing professional will not make deliberately false or deceptive claims about a system or system design, but will instead provide full disclosure of all pertinent system limitations and problems.

A computer professional has a duty to be honest about his or her own qualifications, and about any circumstances that might lead to conflicts of interest.

Membership in volunteer organizations such as ACM may at times place individuals in situations where their statements or actions could be interpreted as carrying the "weight" of a larger group of professionals. An ACM member will exercise care to not misrepresent ACM or positions and policies of ACM or any ACM units.

1.4 Be Fair and Take Action Not to Discriminate

The values of equality, tolerance, respect for others, and the principles of equal justice govern this imperative. Discrimination on the basis of race, sex, religion, age, disability, national origin, or other such factors is an explicit violation of ACM policy and will not be tolerated.

Inequities between different groups of people may result from the use or misuse of information and technology. In a fair society, all individuals would have equal opportunity to participate in, or benefit from, the use of computer resources regardless of race, sex, religion, age, disability, national origin or other such similar factors. However, these ideals do not justify unauthorized use of computer resources nor do they provide an adequate basis for violation of any other ethical imperatives of this code.

1.5 Honor Property Rights Including Copyrights and Patents

Violation of copyrights, patents, trade secrets and the terms of license agreements is prohibited by law in most circumstances. Even when software is not so protected, such violations are contrary to professional behavior. Copies of software should be made only with proper authorization. Unauthorized duplication of materials must not be condoned.

1.6 Give Proper Credit for Intellectual Property

Computing professionals are obligated to protect the integrity of intellectual property. Specifically, one must not take credit for other's ideas or work, even in cases where the work has not been explicitly protected by copyright, patent, etc.

1.7 Respect the Privacy of Others

Computing and communication technology enables the collection and exchange of personal information on a scale unprecedented in the history of civilization. Thus there is increased potential for violating the privacy of individuals and groups. It is the responsibility of professionals to maintain the privacy and integrity of data describing individuals. This includes taking precautions to ensure the accuracy of data, as well as protecting it from unauthorized access or accidental disclosure to inappropriate individuals. Furthermore, procedures must be established to allow individuals to review their records and correct inaccuracies.

This imperative implies that only the necessary amount of personal information be collected in a system, that retention and disposal periods for that information be clearly defined and enforced, and that personal information gathered for a specific purpose not be used for other purposes without consent of the individual(s). These principles apply to electronic communications, including electronic mail, and prohibit procedures that capture or monitor electronic user data, including messages, without the permission of users or bona fide authorization related to system operation and maintenance. User data observed during the normal duties of system operation and maintenance must be treated with strictest confidentiality, except in cases where it is evidence for the violation of law, organizational regulations, or this Code. In these cases, the nature or contents of that information must be disclosed only to proper authorities.

1.8 Honor Confidentiality

The principle of honesty extends to issues of confidentiality of information whenever one has made an explicit promise to honor confidentiality or, implicitly, when private information not directly related to the performance of one's duties becomes available. The ethical concern is to respect all obligations of confidentiality to employers, clients, and users unless discharged from such obligations by requirements of the law or other principles of this Code.

2. MORE SPECIFIC PROFESSIONAL RESPONSIBILITIES

As an ACM computing professional I will ....

2.1 Strive to Achieve the Highest Quality, Effectiveness, and Dignity in Both the Process and Products of Professional Work

Excellence is perhaps the most important obligation of a professional. The computing professional must strive to achieve quality and to be cognizant of the serious negative consequences that may result from poor quality in a system.

2.2 Acquire and Maintain Professional Competence

Excellence depends on individuals who take responsibility for acquiring and maintaining professional competence. A professional must participate in setting standards for appropriate levels of competence, and strive to achieve those standards. Upgrading technical knowledge and competence can be achieved in several ways: doing independent study; attending seminars, conferences, or courses; and being involved in professional organizations.

2.3 Know and Respect Existing Laws Pertaining to Professional Work

ACM members must obey existing local, state, province, national, and international laws unless there is a compelling ethical basis not to do so. Policies and procedures of the organizations in which one participates must also be obeyed. But compliance must be balanced with the recognition that sometimes existing laws and rules may be immoral or inappropriate and, therefore, must be challenged. Violation of a law or regulation may be ethical when that law or rule has inadequate moral basis or when it conflicts with another law judged to be more important. If one decides to violate a law or rule because it is viewed as unethical, or for any other reason, one must fully accept responsibility for one's actions and for the consequences.

2.4 Accept and Provide Appropriate Professional Review

Quality professional work, especially in the computing profession, depends on professional reviewing and critiquing. Whenever appropriate, individual members should seek and utilize peer review as well as provide critical review of the work of others.

2.5 Give Comprehensive and Thorough Evaluations of Computer Systems and Their Impacts, Including Analysis of Possible Risks

Computer professionals must strive to be perceptive, thorough, and objective when evaluating, recommending, and presenting system descriptions and alternatives. Computer professionals are in a position of special trust, and therefore have a special responsibility to provide objective, credible evaluations to employers, clients, users, and the public. When providing evaluations the professional must also identify any relevant conflicts of interest, as stated in imperative 1.3.

As noted in the discussion of principle 1.2 on avoiding harm, any signs of danger from systems must be reported to those who have opportunity and/or responsibility to resolve them. See the guidelines for imperative 1.2 for more details concerning harm, including the reporting of professional violations.

2.6 Honor Contracts, Agreements, and Assigned Responsibilities

Honoring one's commitments is a matter of integrity and honesty. For the computer professional this includes ensuring that system elements perform as intended. Also, when one contracts for work with another party, one has an obligation to keep that party properly informed about progress toward completing that work.

A computing professional has a responsibility to request a change in any assignment that he or she feels cannot be completed as defined. Only after serious consideration and with full disclosure of risks and concerns to the employer or client, should one accept the assignment. The major underlying principle here is the obligation to accept personal accountability for professional work. On some occasions other ethical principles may take greater priority.

A judgment that a specific assignment should not be performed may not be accepted. Having clearly identified one's concerns and reasons for that judgment, but failing to procure a change in that assignment, one may yet be obligated, by contract or by law, to proceed as directed. The computing professional's ethical judgment should be the final guide in deciding whether or not to proceed. Regardless of the decision, one must accept the responsibility for the consequences. However, performing assignments "against one's own judgment" does not relieve the professional of responsibility for any negative consequences.

2.7 Improve Public Understanding of Computing and Its Consequences

Computing professionals have a responsibility to share technical knowledge with the public by encouraging understanding of computing, including the impacts of computer systems and their limitations. This imperative implies an obligation to counter any false views related to computing.

2.8 Access Computing and Communication Resources Only When Authorized to Do So

Theft or destruction of tangible and electronic property is prohibited by imperative 1.2 Avoid Harm to Others. Trespassing and unauthorized use of a computer or communication system is addressed by this imperative. Trespassing includes accessing communication networks and computer systems, or accounts and/or files associated with those systems, without explicit authorization to do so. Individuals and organizations have the right to restrict access to their systems so long as they do not violate the discrimination principle (see 1.4). No one should enter or use another's computer system, software, or data files without permission. One must always have appropriate approval before using system resources, including communication ports, file space, other system peripherals, and computer time.

3. ORGANIZATIONAL LEADERSHIP IMPERATIVES

As an ACM member and an organizational leader, I will ....

BACKGROUND NOTE: This section draws extensively from the draft IFIP Code of Ethics, especially its sections on organizational ethics and international concerns. The ethical obligations of organizations tend to be neglected in most codes of professional conduct, perhaps because these codes are written from the perspective of the individual member. This dilemma is addressed by stating these imperatives from the perspective of the organizational leader. In this context "leader" is viewed as any organizational member who has leadership or educational responsibilities. These imperatives generally may apply to organizations as well as their leaders. In this context "organizations" are corporations, government agencies, and other "employers," as well as volunteer professional organizations.

3.1 Articulate Social Responsibilities of Members of an Organizational Unit and Encourage Full Acceptance of Those Responsibilities

Because organizations of all kinds have impacts on the public, they must accept responsibilities to society. Organizational procedures and attitudes oriented toward quality and the welfare of society will reduce harm to members of the public, thereby serving public interest and fulfilling social responsibility. Therefore, organizational leaders must encourage full participation in meeting social responsibilities as well as quality performance.

3.2 Manage Personnel and Resources to Design and Build Information Systems That Enhance the Quality of Working Life

Organizational leaders are responsible for ensuring that computer systems enhance, not degrade, the quality of working life. When implementing a computer system, organizations must consider the personal and professional development, physical safety, and human dignity of all workers. Appropriate human-computer ergonomic standards should be considered in system design and in the workplace.

3.3 Acknowledge and Support Proper and Authorized Uses of an Organization's Computing and Communication Resources

Because computer systems can become tools to harm as well as to benefit an organization, the leadership has the responsibility to clearly define appropriate and inappropriate uses of organizational computing resources. While the number and scope of such rules should be minimal, they should be fully enforced when established.

3.4 Ensure that Users and Those Who Will Be Affected by a System Have Their Needs Clearly Articulated During the Assessment and Design of Requirements; Later the System Must Be Validated to Meet Requirements

Current system users, potential users and other persons whose lives may be affected by a system must have their needs assessed and incorporated in the statement of requirements. System validation should ensure compliance with those requirements.

3.5 Articulate and Support Policies That Protect the Dignity of Users and Others Affected by a Computing System

Designing or implementing systems that deliberately or inadvertently demean individuals or groups is ethically unacceptable. Computer professionals who are in decision making positions should verify that systems are designed and implemented to protect personal privacy and enhance personal dignity.

3.6 Create Opportunities for Members of the Organization to Learn the Principles and Limitations of Computer Systems

This complements the imperative on public understanding (2.7). Educational opportunities are essential to facilitate optimal participation of all organizational members. Opportunities must be available to all members to help them improve their knowledge and skills in computing, including courses that familiarize them with the consequences and limitations of particular types of systems. In particular, professionals must be made aware of the dangers of building systems around over-simplified models, the improbability of anticipating and designing for every possible operating condition, and other issues related to the complexity of this profession.

4. COMPLIANCE WITH THE CODE

As an ACM member I will ....

4.1 Uphold and Promote the Principles of This Code

The future of the computing profession depends on both technical and ethical excellence. Not only is it important for ACM computing professionals to adhere to the principles expressed in this Code, each member should encourage and support adherence by other members.

4.2 Treat Violations of This Code As Inconsistent with Membership in the ACM

Adherence of professionals to a code of ethics is largely a voluntary matter. However, if a member does not follow this code by engaging in gross misconduct, membership in ACM may be terminated.

This Code and the supplemental Guidelines were developed by the Task Force for the Revision of the ACM Code of Ethics and Professional Conduct: Ronald E. Anderson, Chair, Gerald Engel, Donald Gotterbarn, Grace C. Hertlein, Alex Hoffman, Bruce Jawer, Deborah G. Johnson, Doris K. Lidtke, Joyce Currie Little, Dianne Martin, Donn B. Parker, Judith A. Perrolle, and Richard S. Rosenberg. The Task Force was organized by ACM/SIGCAS and funding was provided by the ACM SIG Discretionary Fund. This Code and the supplemental Guidelines were adopted by the ACM Council on October 16, 1992.

Source: ACM Web site at www.acm.org/. Used with permission.

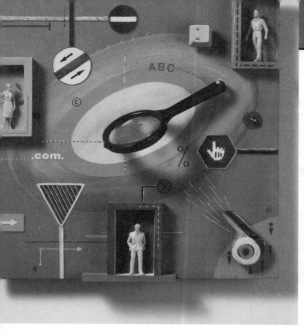

Association of Information Technology Professionals (AITP) Code of Ethics

I acknowledge:

That I have an obligation to management, therefore, I shall promote the understanding of information processing methods and procedures to management using every resource at my command.

That I have an obligation to my fellow members, therefore, I shall uphold the high ideals of AITP as outlined in the Association Bylaws. Further, I shall cooperate with my fellow members and shall treat them with honesty and respect at all times.

That I have an obligation to society and will participate to the best of my ability in the dissemination of knowledge pertaining to the general development and understanding of information processing. Further, I shall not use knowledge of a confidential nature to further my personal interest, nor shall I violate the privacy and confidentiality of information entrusted to me or to which I may gain access.

That I have an obligation to my College or University, therefore, I shall uphold its ethical and moral principles.

That I have an obligation to my employer whose trust I hold, therefore, I shall endeavor to discharge this obligation to the best of my ability, to guard my employer's interests, and to advise him or her wisely and honestly.

That I have an obligation to my country, therefore, in my personal, business, and social contacts, I shall uphold my nation and shall honor the chosen way of life of my fellow citizens.

I accept these obligations as a personal responsibility and as a member of this Association. I shall actively discharge these obligations and I dedicate myself to that end.

AITP Standards of Conduct

These standards expand on the Code of Ethics by providing specific statements of behavior in support of each element of the Code. They are not objectives to be strived for, they are rules that no true professional will violate. It is first of all expected that an information processing professional will abide by the appropriate laws of their country and community. The following standards address tenets that apply to the profession.

In recognition of my obligation to management I shall:

- Keep my personal knowledge up-to-date and insure that proper expertise is available when needed.
- Share my knowledge with others and present factual and objective information to management to the best of my ability.
- Accept full responsibility for work that I perform.
- Not misuse the authority entrusted to me.
- Not misrepresent or withhold information concerning the capabilities of equipment, software or systems.
- Not take advantage of the lack of knowledge or inexperience on the part of others.

In recognition of my obligation to my fellow members and the profession I shall:

- Be honest in all my professional relationships.
- Take appropriate action in regard to any illegal or unethical practices that come to my attention. However, I will bring charges against any person only when I have reasonable basis for believing in the truth of the allegations and without any regard to personal interest.
- Endeavor to share my special knowledge.
- Cooperate with others in achieving understanding and in identifying problems.
- Not use or take credit for the work of others without specific acknowledgement and authorization.
- Not take advantage of the lack of knowledge or inexperience on the part of others for personal gain.

In recognition of my obligation to society I shall:

- Protect the privacy and confidentiality of all information entrusted to me.
- Use my skill and knowledge to inform the public in all areas of my expertise.
- To the best of my ability, insure that the products of my work are used in a socially responsible way.
- Support, respect, and abide by the appropriate local, state, provincial, and federal laws.
- Never misrepresent or withhold information that is germane to a problem or situation of public concern nor will I allow any such known information to remain unchallenged.
- Not use knowledge of a confidential or personal nature in any unauthorized manner or to achieve personal gain.

In recognition of my obligation to my employer I shall:

- Make every effort to ensure that I have the most current knowledge and that the proper expertise is available when needed.
- Avoid conflict of interest and insure that my employer is aware of any potential conflicts.
- Present a fair, honest, and objective viewpoint.
- Protect the proper interests of my employer at all times.
- Protect the privacy and confidentiality of all information entrusted to me.
- Not misrepresent or withhold information that is germane to the situation.
- Not attempt to use the resources of my employer for personal gain or for any purpose without proper approval.
- Not exploit the weakness of a computer system for personal gain or personal satisfaction.

Source: Courtesy of AITP – www.aitp.org

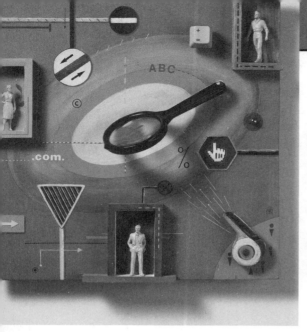

Appendix C

Software Engineering Code of Ethics and Professional Practice

(Version 5.1) as recommended by the
IEEE-CS/ACM Joint Task Force on Software Engineering Ethics and
Professional Practices
Full Version[1]

Preamble

Computers have a central and growing role in commerce, industry, government, medicine, education, entertainment and society at large. Software engineers are those who contribute by direct participation or by teaching, to the analysis, specification, design, development, certification, maintenance and testing of software systems. Because of their roles in developing software systems, software engineers have significant opportunities to do good or cause harm, to enable others to do good or cause harm, or to influence others to do good or cause harm. To ensure, as much as possible, that their efforts will be used for good, software engineers must commit themselves to making software engineering a beneficial and respected profession. In accordance with that commitment, software engineers shall adhere to the following Code of Ethics and Professional Practice.

The Code contains eight Principles related to the behavior of and decisions made by professional software engineers, including practitioners, educators, managers, supervisors and policy makers, as well as trainees and students of the profession. The Principles identify the ethically responsible relationships in which individuals, groups, and organizations participate and the primary obligations within these relationships. The Clauses of each Principle are illustrations of some of the obligations included in these relationships. These obligations are founded in the software engineer's humanity, special care owed to people affected by the work of software engineers, and the unique elements of the practice of software engineering.

The Code prescribes these as obligations of anyone claiming to be or aspiring to be a software engineer.

It is not intended that the individual parts of the Code be used in isolation to justify errors of omission or commission. The list of Principles and Clauses is not exhaustive. The Clauses should not be read as separating the acceptable from the unacceptable in professional conduct in all practical situations. The Code is not a simple ethical algorithm which generates ethical decisions. In some situations standards may be in tension with each other or with standards from other sources. These situations require the software engineer to use ethical judgment to act in a manner which is most consistent with the spirit of the Code of Ethics and Professional Practice, given the circumstances.

Ethical tensions can best be addressed by thoughtful consideration of fundamental principles, rather than blind reliance on detailed regulations. These Principles should influence software engineers to consider broadly who is affected by their work; to examine if they and their colleagues are treating other human beings with due respect; to consider how the public, if reasonably well informed, would view their decisions; to analyze how the least empowered will be affected by their decisions; and to consider whether their acts would be judged worthy of the ideal professional working as a software engineer. In all these judgments concern for the health, safety and welfare of the public is primary; that is, the "Public Interest" is central to this Code.

The dynamic and demanding context of software engineering requires a code that is adaptable and relevant to new situations as they occur. However, even in this generality, the Code provides support for software engineers and managers of software engineers who need to take positive action in a specific case by documenting the ethical stance of the profession. The Code provides an ethical foundation to which individuals within teams and the team as a whole can appeal. The Code helps to define those actions that are ethically improper to request of a software engineer or teams of software engineers.

The Code is not simply for adjudicating the nature of questionable acts; it also has an important educational function. As this Code expresses the consensus of the profession on ethical issues, it is a means to educate both the public and aspiring professionals about the ethical obligations of all software engineers.

PRINCIPLES

Principle 1 PUBLIC Software engineers shall act consistently with the public interest. In particular, software engineers shall, as appropriate:

1.01. Accept full responsibility for their own work.
1.02. Moderate the interests of the software engineer, the employer, the client and the users with the public good.
1.03. Approve software only if they have a well-founded belief that it is safe, meets specifications, passes appropriate tests, and does not diminish quality of life or privacy or harm the environment. The ultimate effect of the work should be to the public good.
1.04. Disclose to appropriate persons or authorities any actual or potential danger to the user, the public, or the environment, that they reasonably believe to be associated with software or related documents.

1.05. Cooperate in efforts to address matters of grave public concern caused by software, its installation, maintenance, support or documentation.

1.06. Be fair and avoid deception in all statements, particularly public ones, concerning software or related documents, methods and tools.

1.07. Consider issues of physical disabilities, allocation of resources, economic disadvantage and other factors that can diminish access to the benefits of software.

1.08. Be encouraged to volunteer professional skills to good causes and contribute to public education concerning the discipline.

Principle 2 CLIENT AND EMPLOYER Software engineers shall act in a manner that is in the best interests of their client and employer and that is consistent with the public interest. In particular, software engineers shall, as appropriate:

2.01. Provide service in their areas of competence, being honest and forthright about any limitations of their experience and education.

2.02. Not knowingly use software that is obtained or retained either illegally or unethically.

2.03. Use the property of a client or employer only in ways properly authorized, and with the client's or employer's knowledge and consent.

2.04. Ensure that any document upon which they rely has been approved, when required, by someone authorized to approve it.

2.05. Keep private any confidential information gained in their professional work, where such confidentiality is consistent with the public interest and consistent with the law.

2.06. Identify, document, collect evidence and report to the client or the employer promptly if, in their opinion, a project is likely to fail, to prove too expensive, to violate intellectual property law, or otherwise to be problematic.

2.07. Identify, document, and report significant issues of social concern, of which they are aware, in software or related documents, to the employer or the client.

2.08. Accept no outside work detrimental to the work they perform for their primary employer.

2.09. Promote no interest adverse to their employer or client, unless a higher ethical concern is being compromised; in that case, inform the employer or another appropriate authority of the ethical concern.

Principle 3 PRODUCT Software engineers shall ensure that their products and related modifications meet the highest professional standards possible. In particular, software engineers shall, as appropriate:

3.01. Strive for high quality, acceptable cost and a reasonable schedule, ensuring significant tradeoffs are clear to and accepted by the employer and the client, and are available for consideration by the user and the public.

3.02. Ensure proper and achievable goals and objectives for any project on which they work or propose.

3.03. Identify, define and address ethical, economic, cultural, legal and environmental issues related to work projects.

3.04. Ensure that they are qualified, by an appropriate combination of education and training, and experience, for any project on which they work or propose to work.

3.05. Ensure an appropriate method is used for any project on which they work or propose to work.

3.06. Work to follow professional standards, when available, that are most appropriate for the task at hand, departing from these only when ethically or technically justified.

3.07. Strive to fully understand the specifications for software on which they work.

3.08. Ensure that specifications for software on which they work have been well documented, satisfy the users' requirements and have the appropriate approvals.

3.09. Ensure realistic quantitative estimates of cost, scheduling, personnel, quality and outcomes on any project on which they work or propose to work and provide an uncertainty assessment of these estimates.

3.10. Ensure adequate testing, debugging, and review of software and related documents on which they work.

3.11. Ensure adequate documentation, including significant problems discovered and solutions adopted, for any project on which they work.

3.12. Work to develop software and related documents that respect the privacy of those who will be affected by that software.

3.13. Be careful to use only accurate data derived by ethical and lawful means, and use it only in ways properly authorized.

3.14. Maintain the integrity of data, being sensitive to outdated or flawed occurrences.

3.15. Treat all forms of software maintenance with the same professionalism as new development.

Principle 4 JUDGMENT Software engineers shall maintain integrity and independence in their professional judgment. In particular, software engineers shall, as appropriate:

4.01. Temper all technical judgments by the need to support and maintain human values.

4.02. Only endorse documents prepared under their supervision or within their areas of competence and with which they are in agreement.

4.03. Maintain professional objectivity with respect to any software or related documents they are asked to evaluate.

4.04. Not engage in deceptive financial practices such as bribery, double billing, or other improper financial practices.

4.05. Disclose to all concerned parties those conflicts of interest that cannot reasonably be avoided or escaped.

4.06. Refuse to participate, as members or advisors, in a private, governmental or professional body concerned with software related issues, in which they, their employers or their clients have undisclosed potential conflicts of interest.

Principle 5 MANAGEMENT Software engineering managers and leaders shall subscribe to and promote an ethical approach to the management of software development and maintenance. In particular, those managing or leading software engineers shall, as appropriate:

5.01. Ensure good management for any project on which they work, including effective procedures for promotion of quality and reduction of risk.

5.02. Ensure that software engineers are informed of standards before being held to them.

5.03. Ensure that software engineers know the employer's policies and procedures for protecting passwords, files and information that is confidential to the employer or confidential to others.

5.04. Assign work only after taking into account appropriate contributions of education and experience tempered with a desire to further that education and experience.

5.05. Ensure realistic quantitative estimates of cost, scheduling, personnel, quality and outcomes on any project on which they work or propose to work, and provide an uncertainty assessment of these estimates.

5.06. Attract potential software engineers only by full and accurate description of the conditions of employment.

5.07. Offer fair and just remuneration.

5.08. Not unjustly prevent someone from taking a position for which that person is suitably qualified.

5.09. Ensure that there is a fair agreement concerning ownership of any software, processes, research, writing, or other intellectual property to which a software engineer has contributed.

5.10. Provide for due process in hearing charges of violation of an employer's policy or of this Code.

5.11. Not ask a software engineer to do anything inconsistent with this Code.

5.12. Not punish anyone for expressing ethical concerns about a project.

Principle 6 PROFESSION Software engineers shall advance the integrity and reputation of the profession consistent with the public interest. In particular, software engineers shall, as appropriate:

6.01. Help develop an organizational environment favorable to acting ethically.

6.02. Promote public knowledge of software engineering.

6.03. Extend software engineering knowledge by appropriate participation in professional organizations, meetings and publications.

6.04. Support, as members of a profession, other software engineers striving to follow this Code.

6.05. Not promote their own interest at the expense of the profession, client or employer.

6.06. Obey all laws governing their work, unless, in exceptional circumstances, such compliance is inconsistent with the public interest.

6.07. Be accurate in stating the characteristics of software on which they work, avoiding not only false claims but also claims that might reasonably be supposed to be speculative, vacuous, deceptive, misleading, or doubtful.

6.08. Take responsibility for detecting, correcting, and reporting errors in software and associated documents on which they work.

6.09. Ensure that clients, employers, and supervisors know of the software engineer's commitment to this Code of ethics, and the subsequent ramifications of such commitment.

6.10. Avoid associations with businesses and organizations which are in conflict with this code.

6.11. Consider that violations of this Code are inconsistent with being a professional software engineer.

6.12. Express concerns to the people involved when significant violations of this Code are detected unless this is impossible, counter-productive, or dangerous.

6.13. Report significant violations of this Code to appropriate authorities when it is clear that consultation with people involved about significant violations of this Code, is impossible, counter-productive or dangerous.

Principle 7 COLLEAGUES Software engineers shall be fair to and supportive of their colleagues. In particular, software engineers shall, as appropriate:

7.01. Encourage colleagues to adhere to this Code.

7.02. Assist colleagues in professional development.

7.03. Credit fully the work of others and refrain from taking undue credit.

7.04. Review the work of others in an objective, candid, and properly-documented way.

7.05. Give a fair hearing to the opinions, concerns, or complaints of a colleague.

7.06. Assist colleagues in being fully aware of current standard work practices including policies and procedures for protecting passwords, files and other confidential information, and security measures in general.

7.07. Not unfairly intervene in the career of any colleague; however, concern for the employer, the client or public interest may compel software engineers, in good faith, to question the competence of a colleague.

7.08. In situations outside of their own areas of competence, call upon the opinions of other professionals who have competence in that area.

Principle 8 SELF Software engineers shall participate in lifelong learning regarding the practice of their profession and promote an ethical approach to the practice of the profession. In particular, software engineers shall continually endeavor to:

8.01. Further their knowledge of developments in the analysis, specification, design, development, maintenance and testing of software and related documents, together with the management of the development process.

8.02. Improve their ability to create safe, reliable, and useful quality software at reasonable cost and within a reasonable time.

8.03. Improve their ability to produce accurate, informative, and well-written documentation.

8.04. Improve their understanding of the software and related documents on which they work and of the environment in which they will be used.

8.05. Improve their knowledge of relevant standards and the law governing the software and related documents on which they work.

8.06. Improve their knowledge of this Code, its interpretation, and its application to their work.

8.07. Not give unfair treatment to anyone because of any irrelevant prejudices.

8.08. Not influence others to undertake any action that involves a breach of this Code.

8.09. Consider that personal violations of this Code are inconsistent with being a professional software engineer.

This Code was developed by the IEEE-CS/ACM joint task force on Software Engineering Ethics and Professional Practices (SEEPP):

Executive Committee:

Donald Gotterbarn (Chair), Keith Miller and Simon Rogerson;

Members:

Steve Barber, Peter Barnes, Ilene Burnstein, Michael Davis, Amr El-Kadi, N. Ben Fairweather, Milton Fulghum, N. Jayaram, Tom Jewett, Mark Kanko, Ernie Kallman, Duncan Langford, Joyce Currie Little, Ed Mechler, Manuel J. Norman, Douglas Phillips, Peter Ron Prinzivalli, Patrick Sullivan, John Weckert, Vivian Weil, S. Weisband and Laurie Honour Werth.

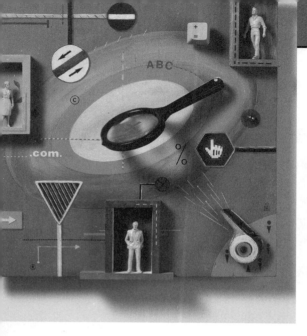

Institute of Electrical and Electronics Engineers (IEEE) Code of Ethics

We, the members of the IEEE, in recognition of the importance of our technologies in affecting the quality of life throughout the world, and in accepting a personal obligation to our profession, its members and the communities we serve, do hereby commit ourselves to the highest ethical and professional conduct and agree:

1. To accept responsibility in making engineering decisions consistent with the safety, health and welfare of the public, and to disclose promptly factors that might endanger the public or the environment;
2. To avoid real or perceived conflicts of interest whenever possible, and to disclose them to affected parties when they do exist;
3. To be honest and realistic in stating claims or estimates based on available data;
4. To reject bribery in all its forms;
5. To improve the understanding of technology, its appropriate application, and potential consequences;
6. To maintain and improve our technical competence and to undertake technological tasks for others only if qualified by training or experience, or after full disclosure of pertinent limitations;
7. To seek, accept, and offer honest criticism of technical work, to acknowledge and correct errors, and to credit properly the contributions of others;
8. To treat fairly all persons regardless of such factors as race, religion, gender, disability, age, or national origin;
9. To avoid injuring others, their property, reputation, or employment by false or malicious action;
10. To assist colleagues and co-workers in their professional development and to support them in following this code of ethics.

Approved by the IEEE Board of Directors
August 1990

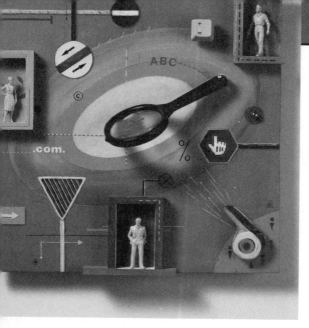

GLOSSARY

A

active click wrap license A software license transmitted electronically and activated the instant the user installs the software or accesses the information.

affiliated Web sites A collection of Web sites served by a single advertising network.

anonymous expression The ability to state one's opinions without revealing one's identity.

anonymous remailer An Internet service that allows an e-mail sender to remain anonymous by using a computer to strip the originating address from an e-mail message and then forward it to its intended recipient.

B

black-box testing Testing of software as a unit that has expected input and output behaviors, with the inner workings unknown.

breach of the duty of care The failure to act as a reasonable person would act.

breach of warranty A version of product liability where the product fails to meet its warranty.

bribery The paying of money, property, favors, or anything else of value to someone in business or government in order to obtain a business advantage.

business information system A set of interrelated components including hardware, software, databases, networks, people, and procedures that collect data, process it, and disseminate the output.

C

Capability Maturity Model® for Software A model that defines six levels of software development process maturity and identifies the issues most critical to software quality and process improvement.

Carnivore A highly controversial system used by the FBI to monitor selected e-mail messages.

certification A process administered by a profession or other organization that one undertakes voluntarily to prove competency in a set of skills.

chargebacks Disputed credit card transactions.

chief privacy officer (CPO) A senior manager responsible for training employees about privacy, checking the company's privacy policies for potential risks and then figuring out if and how to fill gaps, and managing a customer privacy dispute and verification process.

click stream data Data that allows the monitoring of a person's movements throughout any affiliated Web sites.

click-on license A software license transmitted electronically and activated the instant the user installs the software or accesses the information.

code of conduct A set of formal, written statements about the purpose of an organization, its values, and the principles that guide its employees' actions.

co-employment A relationship between two or more employers in which each has actual or potential legal rights and duties with respect to the same employee or group of employees.

collaborative filtering A form of personalization software that offers consumer recommendations based on the types of products purchased by other individuals who bought the same product as another customer.

collusion Cooperation between an employee and an outsider to commit fraud.

common good approach An approach to ethical decision making that is based on a vision of society as a community whose members work together to achieve a common set of values and goals.

compiler A language translator that converts computer program statements expressed in a source language (e.g., COBOL, Pascal, C) into machine language (a series of binary codes, 0s and 1s) that the computer can execute.

competitive intelligence The gathering of legally obtainable information that will help a company gain an advantage over its rivals.

contextual commerce A form of personalization software that associates product promotions and other e-commerce offerings with specific content a user may be receiving in a news story online.

contributory negligence A defense in a negligence case where the defendant proves that the plaintiffs' own actions contributed at least in part to their injuries.

contingent workforce Independent contractors; individuals brought in through employment agencies; on-call or day laborers; and workers on site whose services are provided by contract firms, such as outsourced information technology workers.

cookie A unique identifier that passes information back to a marketer's computer when one surfs the Web.

235

copyright A form of protection that grants the authors the exclusive right to distribute, display, perform, or reproduce the work in copies or to prepare derivative works based upon the work.

corporate ethics officer A senior-level manager responsible for establishing policies and training to improve the ethical behavior of the members of an organization.

cracker A hacker who performs illegal acts such as breaking into other people's networks and systems, defacing Web pages, crashing computers, spreading harmful programs or hateful messages, or writing scripts and automatic programs that let other people do these things.

criminal fraud The obtaining of title to property through deception or trickery.

cybersquatter Someone who registers a Web site domain containing a famous trademark or company name to which the person has no connection, with the hope that the trademark's legitimate owners will pay to gain ownership of the domain.

cyberterrorist Someone who intimidates or coerces a government to further a political or social objective by launching attacks or threatening attacks against computers, networks, and the information stored therein.

cryptography The science of encoding messages so that only the sender and the intender receiver can understand them.

D

decision support system A type of business information system used to improve decision-making effectiveness.

decompiler Software tool that can read machine language and produce the source code.

defamation The publication of alleged fact which is false and which harms another.

definitions Virus detection information used to update anti-virus software.

deliverables Products (e.g., a statement of requirements, flowcharts, user documentation) developed during the various stages of the software development process.

demographic filtering A form of personalization software that augments click-stream data and user-supplied data with demographic information associated with user zip codes to make product suggestions.

denial-of-service attack An attack in which a malicious hacker takes over computers on the Internet and causes them to flood a target site with demands for data and other small tasks.

duty of care The obligation we all owe each other—the duty not to cause any unreasonable harm or risk of harm.

dynamic testing The testing of software by entering test data and comparing the actual results to the expected results.

E

egress filtering A process by which corporations ensure that packets with false return addresses do not leave their corporate network.

electronic signature Information or data attached or logically associated with an electronic record and adopted by the person with the intent to sign an agreement.

employee leasing The placement by an employer of all or most of its existing work force onto the payroll of an employee leasing firm in an explicit co-employment relationship; doing so "outsources" the administration of all payroll, benefits, and other human resources activities.

encryption The process of converting an original message into a form that can be understood only by the intended recipients.

ethics An individual's personal beliefs regarding right and wrong behavior.

F

Failure Mode and Effects Analysis (FMEA) A reliability evaluation technique used to determine the effect of system and equipment failures.

fair use doctrine Sets forth four factors for courts to consider in determining whether a particular use of copyrighted property is a fair use and can be allowed without penalty.

fairness approach An approach to ethical decision making that focuses on how fairly our actions and policies distribute benefits and burdens among those affected by the decision.

False Claims Act A federal law enacted in 1863 that was designed to entice whistleblowers to come forward by offering them a share of the money recovered.

firewall A hardware and/or software device that serves as a barrier between a company and the outside world, limiting access into and out of the company's network based on its IT usage policy.

Foreign Corrupt Practices Act Makes it a crime to bribe a foreign official, a

foreign political party official, or a candidate for foreign political office.

fraud Theft with deception.

G

get data Data gathered as one browses the Web; identifies the sites visited and information requested.

H

H-1B worker A foreign person working in the US with an H-1B temporary working visa for people who work in specialty occupations: jobs that require a four-year bachelor's degree or higher in a specific field, or the equivalent experience.

hacker An individual who tests the limitations of systems out of intellectual curiosity, trying to see what they access and how far can they go.

honeypot A computer on a network that contains no data or applications critical to the company, but that has enough interesting data to lure an intruder so that they can be observed in action.

I

industrial espionage The use of illegal means to obtain business information not available to the general public.

industrial spies Hackers who use illegal means to obtain trade secrets about the competitors of the firm for which they are hired.

ingress filtering A process by which Internet Service Providers prevent incoming packets with false IP addresses from being passed on.

237

integration testing Software testing that follows successful unit testing, where the software units are combined into an integrated subsystem that undergoes rigorous testing.

integrity Used to refer to people who act in ways that are consistent with their beliefs.

intellectual property Includes works of the mind such as art, books, films, formulae, inventions, music, and processes which are distinct somehow and are "owned" and/or created by a single entity.

intentional misrepresentation Occurs when a seller or lessor knowingly either misrepresents the quality of a product or conceals a defect in it.

Internet filter Software that can be installed on a personal computer along with a Web browser to block access to certain Web sites that contain inappropriate or offensive material.

intrusion detection system Monitors system and network resources and activities and, using information gathered from these sources, notifies the authorities when it identifies a possible intrusion.

ISO 9000 A series of standards that require organizations to develop formal quality management systems that focus on identifying and meeting the needs, wants, and expectations of their customers.

J

John Doe lawsuit A lawsuit where the true identity of the defendant is temporarily unknown.

K

key A variable value that is applied using an algorithm to a string or block of text to produce encrypted text or to decrypt encrypted text.

L

lamer A technically inept hacker.

libel A written statement of alleged fact which is false and which harms another person.

licensing A process generally administered at the state level in the United States which a professional must undertake to prove that they can practice their profession in a manner that is ethical and safe to the general public.

logic bomb A type of Trojan horse that executes when specific conditions occur.

M

morality Social conventions about right and wrong human conduct that are so widely shared that they are the basis for an established common consensus.

N

negligence Failure to do something which a reasonable person would do, or doing something which a prudent and reasonable person would not do.

non-compete agreements An agreement between an employer and employee that requires employees to not work for any competitors for a period of time after leaving the original employer.

238

nondisclosure clause A clause in an employment contract that requires employees to refrain from revealing secrets that they learn at work.

O

offshore outsourcing A variation of outsourcing where the work is done by an organization using employees who perform the work while in a foreign country.

opt-in An approach to data collection that requires the data collector to get specific permission from a consumer before collecting any of his or her data.

opt-out An approach to data collection that assumes that consumers approve of companies collecting and storing their personal information and requires consumers to specifically tell companies, one by one, not to collect data about them.

outsourcing When an organization contracts with a firm that has expertise in operating a specific client function to perform that function on an on-going basis.

P

patent A form of legal protection of an invention that enables the inventor to take legal action against those who, without the inventor's permission, manufacture, use, or sell the invention during the period of time the patent is in force.

personalization software Software used by marketers to optimize the number, frequency, and mixture of their ad placements, and to test how visitors react to new ads.

platform for privacy preferences (P3P) Screening technology being proposed to shield users from sites that don't provide the level of privacy protection they desire.

prior art The existing body of knowledge that is available at a given time to a person or ordinary skill in the art. Prior art is searched as part of the patent application process; the patent cannot be issued for an invention whose professed improvements are already present in, or are obvious from, the prior art.

private key encryption A system that uses only one key to both encode and decode a message.

post data Data entered into blank fields on a Web site when a consumer signs up for a service.

product liability The liability of manufacturers, sellers, lessors, and others for injuries caused by defective products.

profession A calling requiring specialized knowledge and often long and intensive academic preparation, whose work requires the consistent exercise of discretion and judgment on its performance, and where the nature of the work is predominately intellectual and varied in character and the output produced or the result accomplished cannot be standardized in relation to a given period of time.

professional code of ethics Statement of the principles and core values essential to the work of a particular occupational group.

professional malpractice The breaching of one's professional duty of care, which causes the professional to be liable for the injury his or her negligence causes.

239

public key encryption An encryption system that uses a public key to encode a message and a private key to decode messages.

Q

quality management The definition, measurement, and quality refinement of the information systems development process and the products developed during the various stages of the process.

qui tam A provision of the Federal False Claims Act that allows private citizens to file a lawsuit in the name of the U.S. Government charging fraud by government contractor and others who receive or use government funds, and to share in any money recovered.

R

reasonable assurance An approach to security in which a manager uses his or her judgment to ensure that the cost of control does not exceed the benefit to be obtained by implementation or the possible risk involved.

reasonable person standard How an objective, careful, and conscientious person would act under certain circumstances.

reasonable professional standard How a trained and experienced professional would act under certain circumstances.

redundancy The provision of multiple interchangeable hardware components to perform a single function in order to cope with failures and errors.

reliability The probability of a component or system performing its mission over a certain length of time.

resume inflation Lying on one's resume by saying that one is competent in an IT skill that is in high demand when one does not actually have such competency.

reverse engineering The process of analysis of an existing software system to create a new representation of the system in a different form or at a higher level of abstraction.

risk The product of the likelihood of a negative event happening times the impact of such an event happening.

risk assessment An organization's review of potential threats to its computers and network and the probability of those threats occurring.

rules-based personalization software Uses business rules tied to customer-provided preference information or online behavior to determine the most appropriate Web page views and product information to display.

S

safety critical system System whose failure may cause injury or death to human beings.

script kiddie A technically inept hacker.

security policy Defines the security requirements of an organization and describes the controls and sanctions to be used to meet those requirements.

shrink-wrap license A software license that accompanies the disk or package containing the software program or information.

slander An oral statement of alleged fact which is false and which harms another person.

smart card Credit card that contains a memory chip that is updated with encrypted data every time the card is read.

social audit Studies that identify corporate ethical lapses committed in the past and set directives for avoiding similar missteps in the future.

software defect Any error that, if not removed, would cause a system to fail to meet the needs of its users.

software development methodology A standard, proven work process that enables systems analysts, programmers, project managers, and others to make controlled and orderly progress in developing high-quality software.

software piracy The act of illegally making copies of software or enabling others to access software to which they are not entitled.

software quality The degree to which the attributes of a software product enable it to meet the needs of its users.

software quality assurance Those methods within the software development methodology that are used to guarantee that the software being developed will operate reliably.

spamming The sending of many copies of the same message in an attempt to force a large number of people to read a message they would otherwise choose not to receive.

spoofing The programming of computers to put false return addresses on the packets they send out.

stakeholder Someone who stands to gain or lose from how a particular situation is resolved.

static testing The use of special software to look for suspicious patterns in programs that might indicate a software defect.

strict liability A version of product liability where the defendant is responsible for injuring another person regardless of negligence or intent.

system testing A form of software testing that follows successful integration testing where the various subsystems are combined and testing is conducted to test the entire system as a complete entity.

T

trade secret A piece of information used in a business that the company has taken strong measures to keep confidential, represents something of economic value, required effort or cost to develop, and has some degree of uniqueness or novelty.

trademark Something (such as a logo, package design, phrase, sound, or word) that helps a consumer distinguish one company's products from another's.

Trojan horse A program that gets secretly installed on a computer, planting a harmful payload that can allow the hacker to do such things as steal passwords or spy on users by recording keystrokes and transmitting them to a third party.

241

U

utilitarianism approach An approach to ethical decision making that states that when we have a choice between alternative actions or social policies, we choose the one that has the best overall consequence for all persons directly or indirectly affected.

user acceptance testing An independent test performed by trained end users to ensure that a system operates as expected from their viewpoint.

V

value system The complex scheme of moral values that one elects to live by.

vice A moral habit that inclines one to do what is generally unacceptable to society.

virtue A moral habit that inclines one to do what is generally acceptable to society.

virtue ethics approach A philosophical approach to ethical decision making that suggests that, when faced with a complex ethical dilemma, people simply do what they are most comfortable doing or do what they think a person they admire would do. The assumption is that people will be guided by their virtues to reach the "right" decision.

virus A computer program that attaches to a file and replicates itself repeatedly, typically without user knowledge or permission.

virus signature A specific sequence of bytes that identifies a virus.

W

warranty An assurance to buyers or lessees that a product meets certain standards of quality.

whistle blowing An effort by an employee of a company to attract the attention of others to a negligent, illegal, unethical, abusive, or dangerous act by the company that threatens the public interest.

white-box testing Testing of software as a unit that has expected input and output behaviors and whose internal workings are known. Involves the testing of all possible logic paths through the software unit and is done with thorough knowledge of the logic of the software unit.

worms Harmful computer programs that differ from viruses because they have the capability to self-propagate without human intervention.

Z

zombies Computers taken over by a hacker launching a denial-of-service attack and commanded to send repeated requests for access to a single target site.

Index

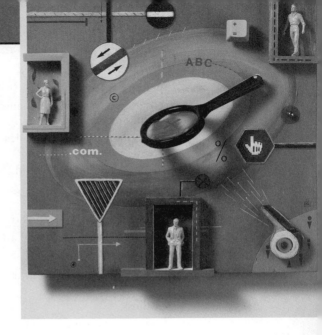

B

Backup processes, 70

BBB online. *see* Better Business Bureau Online (BBB online)

Benchmark evaluation, codes of ethics and, 34

Better Business Bureau Online (BBB online), 90

Bill Gates, 162–163

Borland, Lotus *vs.*, 158–160

Breach of the duty of care, 40

Bribery, 7–8

Brown & Williamson Tobacco Company, 207–208

Business
ethics, fostering of, 6–12
ethics risk and, 5
good *vs.* bad, 8–9
success and ethics, 12–13
Utilitarian approach and, 16
virtue ethics and, 16

Business information systems, 165–166

Business organizations, John Doe lawsuit case and, 121

C

CALEA. *see* Communications Assistance for Law Enforcement Act (CALEA)

Capability Maturity Model for Software (SW-CMM), 172

Carnegie Mellon University, Software Engineering Institute, 54

Carnivore, 101

Catalytic Software, 208–209

CCP. *see* Certified Computing Professional (CCP)

CDA. *see* Communications Decency Act (CDA)

CERT/CC. *see* Computer Emergency Response Team Coordination Center (CERT/CC)

Certification, 35
industry association, 36–37
Microsoft certified systems engineer (MCSE), 36
process, 36
vendor, 36

Certified Computing Professional (CCP), 36, 37

Chargebacks, 63

Charles Schwab & Co., 48

Chief Information Officer (CIO), 28. *see also* IT professionals

Chief Privacy Officer (CPO), 97

Children Online Privacy Protection Act (COPA), 89

Children's Internet Protection Act of 2000, 117–119

CIO. *see* Chief Information Officer (CIO)

Clean Air Act, 201

Clean Water Act, 201

Client, IT professional ethics and, 30–31

Code of conduct, 10–11, 12

Code of ethics
Association for Computing Machinery (ACM) and, 34–35, 212–219
Association of Information Technology Professionals (AITP) and, 35, 220–222
benchmark evaluation and, 34
Computer Society of the Institute of Electrical and Electronics Engineers and, 35
Institute of Electrical and Electronics Engineers and, 232–233
software engineering and, 224–240

Code of Fair Information Practices, 97

Co-employment relationship, 191

Collaborative filtering, 95

Collusion, 61

Commercial software and security, 56

Common good approach, 17

247